International Conference on CRISPR Technologies 2017

Raleigh, North Carolina, USA
4-6 December 2017

ISBN: 978-1-7138-2266-0

Printed from e-media with permission by:

Curran Associates, Inc.
57 Morehouse Lane
Red Hook, NY 12571

Some format issues inherent in the e-media version may also appear in this print version.

Printed with permission by Curran Associates, Inc. (2021)

For permission requests, please contact AIChE
at the address below.

AIChE
120 Wall Street, FL 23
New York, NY 10005-4020

Phone: (800) 242-4363
Fax: (203) 775-5177

www.aiche.org

Additional copies of this publication are available from:

Curran Associates, Inc.
57 Morehouse Lane
Red Hook, NY 12571 USA
Phone: 845-758-0400
Fax: 845-758-2633
Email: curran@proceedings.com
Web: www.proceedings.com

TABLE OF CONTENTS

KEYNOTES

GENOME EDITING AND GENE REGULATION IN HUMAN HEALTH

SESSION 2

SESSION 3

ACHIEVING EFFICIENT DELIVERY AND EDITING

CRISPR TECHNOLOGIES BEYOND GENOME EDITING AND GENE REGULATION

GENOME EDITING AND GENE REGULATION IN AGRICULTURE

HORIZONS OF CRISPR BIOLOGY

POSTER SESSION

Genome and Epigenome Editing with CRISPR/Cas9 for Gene Therapy and Cell Programming

Conference: International Conference on CRISPR Technologies

Year: 2017

Proceeding: International Conference on CRISPR Technologies

Group: General Submissions

Session: Keynotes

Time: Monday, December 4, 2017 - 2:05am-2:30am

Authors:

Gersbach, C. A., Duke University

The advent of genome engineering technologies, including the RNA-guided CRISPR/Cas9 system, has enabled the precise editing and regulation of endogenous human genes and epigenetic states. We have applied these tools to the correction of mutations that cause genetic disease and also adapted them to manipulate the epigenome and control cell fate decisions. For example, we engineered CRISPR/Cas9-based nucleases to correct the human dystrophin gene that is mutated in Duchenne muscular dystrophy patients. When we delivered these nucleases to cells from patients with this disease, the correct gene reading frame and expression of the functional dystrophin protein were restored*in vitro* and following cell transplantation into mouse models *in vivo*. When delivered directly to a mouse model of this disease, gene editing by the CRISPR/Cas9 system led to gene restoration and improvement of biochemical and mechanical muscle function. More recently, we have developed and tested a novel humanized mouse model of Duchenne muscular dystrophy for preclinical development of gene editing therapies. In other studies, we have engineered CRISPR/Cas9-based tools to regulate the expression of endogenous genes and applied these tools to control diverse genes relevant to disease, development, and differentiation. Genome-wide analysis of the DNA-binding, gene regulation, and chromatin remodeling by these targeted epigenome modifiers has demonstrated their exceptional specificity. We have recently applied these technologies to control the decisions of stem cells to become specific cell fates and reprogram cell types into other lineages that could be used for drug screening and disease modeling. Additionally, we are developing epigenome editing tools to annotate the function of the non-coding genome and illuminate the role of gene regulation in disease susceptibility and drug response. Collectively, these studies demonstrate the potential of modern genome engineering technologies to capitalize on the products of the Genomic Revolution and transform medicine, science, and biotechnology.

Discovery of Novel CRISPR-Cas Systems and Implications for the Evolution of Microbial Adaptive Immunity

Conference: International Conference on CRISPR Technologies

Year: 2017

Proceeding: International Conference on CRISPR Technologies

Group: General Submissions

Session: Keynotes

Time: Monday, December 4, 2017 - 1:40am-2:05am

Authors:

Koonin, E. V., NIH/NLM/NCBI

Class 2 CRISPR-Cas systems are characterized by effector modules that consist of a single multidomain protein, such as Cas9 or Cpf1. We designed a computational pipeline for the discovery of novel class 2 variants and used it to identify six new CRISPR-Cas subtypes. The diverse properties of these new systems provide potential for the development of versatile tools for genome editing and regulation, and several such tools have already been developed starting with the predictions produced using our pipeline. I will present a comprehensive census of Class 2 CRISPR-Cas types and subtypes in complete and draft bacterial and archaeal genomes and outline evolutionary scenarios for the independent origin of different class 2 CRISPR-Cas systems from mobile genetic elements, along with a general concept of evolution of microbial adaptive immunity.

RNA Secondary Structure Increases the Specificity of Class 2 CRISPR Effectors

Conference: International Conference on CRISPR Technologies

Year: 2017

Proceeding: International Conference on CRISPR Technologies

Group: General Submissions

Session: Genome editing and gene regulation in human health

Time: Monday, December 4, 2017 - 2:05am-2:30am

Authors:

Kocak, D. D., Duke University
Josephs, E. A., Duke University
Kwon, J. B., Duke University
Adkar, S. S., Duke University
Fitzgibbon, C. J., Duke University
Klann, T. S., Duke University
Thakore, P. I., Duke University
Marszalek, P. E., Duke University
Gersbach, C. A., Duke University

CRISPR systems have been widely adopted for both therapeutic applications and biological studies. In some cases, these bacterial nucleases have demonstrated off-target activity when applied in mammalian cells. This creates a potential hazard for therapeutic applications and could possibly confound results in biological research. Therefore, improving the specificity of these nucleases is of broad interest. Here we show that guide RNAs (gRNAs) with engineered secondary structure in the targeting region (hairpin-gRNAs) can have increased specificity by several orders of magnitude when combined with various CRISPR effectors. Focusing first on Cas9 from S. pyogenes(SpCas9), we show that rationally designed hp-gRNAs form in human cells and tune Cas9 binding. We then show that hp-gRNAs increase the specificity of gene editing using five different CRISPR endonucleases: (i)SpCas9, (ii) Cas9 from S. aureus (SaCas9), (iii) an SaCas9 variant with relaxed PAM specificity (SaCas9-KKH), (iv) Cpf1 from Acidaminococcussp.BV3L6 (AsCpf1), and (v) Cpf1 from Lachnospiraceae bacterium ND2006 (LbCpf1). Our results suggest RNA secondary structure as a fundamental parameter that can tune the activity of class 2 CRISPR systems.

Gene Repression Via CRISPR-Assisted Modification of Specific Chromatin Marks

Conference: International Conference on CRISPR Technologies

Year: 2017

Proceeding: International Conference on CRISPR Technologies

Group: General Submissions

Session: Genome editing and gene regulation in human health

Time: Monday, December 4, 2017 - 1:15am-1:40am

Authors:

Holtzman, L., Duke University
Rathinavelu, J., Duke University
Gersbach, C. A., Duke University

The CRISPR-CAS9 system allows for efficient and specific nuclease targeting, simply by programming target site recognition with a short guide RNA. Fusions of its deactivated variant (dCas9) are being used widely to activate and repress gene expression in a highly specific and robust manner when targeted to gene promoters and enhancers. For example, dCas9 fused to either the VP64 activation domain, the KRAB repression domain, or the P300 acetyltransferase domain have been shown to alter the modification of chromatin marks H3K4, H3K9, and H3K27 in a context-dependent manner. In this current study, we developed fusions of dCas9 to other various chromatin modifiers (SETDB2, SIRT6, EHMT2) and demonstrated their transcriptional repression of the genes MeCP2, FMR1 and SOX2. Additionally, we've characterized the chromatin changes that those fusions deposit. In particular, the Sirt6 fusion removes acetyl marks from H3K9 and H4K16 residues, while the SETDB2 fusion catalyzes targeted tri-methylation of the H3K9 residue, but does not affect H3K27 marks. The latter contrasts with the more commonly used KRAB-based repressor, which recruits a repressor complex that leads to accumulation of a variety of repressive histone marks. Ongoing studies include exploring the combinatorial effect of depositing multiple specific chromatin marks at the same locus and how this translates to changes in gene expression. These fusions expand the epigenome-engineering toolkit and will facilitate more specific manipulation of chromatin and understanding its role in gene regulation, which ultimately may also enable more deterministic programming of gene expression.

Fluorescent Guide RNAs for Understanding Grnas Expressed from Pol II Promoters

Conference: International Conference on CRISPR Technologies

Year: 2017

Proceeding: International Conference on CRISPR Technologies

Group: General Submissions

Session: Genome editing and gene regulation in human health

Time: Monday, December 4, 2017 - 12:25pm-12:50pm

Authors:

Menn, D., Arizona State University
Ebrahimkhani, M., Mayo Clinic College of Medicine and Science
Wang, X., Arizona State University
Kiani, S., Arizona State University

Introduction: CRISPR technology has quickly become an important tool in the synthetic biology repertoire, used for applications such as induced mutation and gene silencing, genome labeling, and activation/repression. The relatively straight-forward design principles of the gRNAs, together with the ease with which gRNAs or Cas9 proteins can be modified to recruit or contain other active domains, promises to simplify and expand the field of de novo genetic circuits and allow researchers convenient ways to interact with existing networks. This shows great potential for bioproduction and therapeutic applications, either rewiring faulty genetic networks or normalizing unbalanced systems. However, most systems of interest are composed primarily of RNA Polymerase II (Pol II) driven components, while non-coding gRNAs are most commonly produced by RNA Polymerase III (Pol III). While several methods exist to convert the mRNA products of Pol II promoters into functional gRNAs[1], none of these methods produce gRNA capable of inducing downstream effects as powerfully as those produced from a strong Pol III promoter. In this study, we aim to increase the efficiency of Pol II gRNA production in order to allow researchers to interface more dynamically with existing gene networks.

Materials and Methods: Plasmids were made containing either Pol III or inducible Pol II promoters driving gRNAs containing the fluorescent RNA aptamer Broccoli[2]. Pol II constructs included either ribozymes or Csy4 sites, to allow mRNA editing into a functional gRNA. These basic constructs were then modified to increase their effectiveness in producing gRNAs:

modified terminators, intronic additions, multiplexing, and an RNA nuclear localization sequence. Simple repressor circuits, in which the gRNA repressed the expression of infrared fluorescent protein (iRFP) were transfected into human HEK293rtTA cells and analyzed via flow cytometry. Fluorescence increase of the Broccoli aptamer was compared to the decrease of the iRFP signal to determine effectiveness of gRNA repression.

Results and Discussion: Initial data indicates that the gRNA transcript can be modified to fluoresce without affecting its function, and that mRNA can be edited into functional fluorescent guides through multiple methods. Further, we have demonstrated that that Pol II driven repression efficiency can be increased by multiplexing several repeats of the gRNA together, as well as by changing the gene's terminator to one containing a minimal polyA sequence, thereby reducing nuclear export.

Conclusions: This work shows that gRNA expression from Pol II promoters can be improved through several different methods. They can be used in tandem or employed individually for niche situations. One limitation of all of these methods, however, is the efficiency of the editors which cleave mRNA into functional gRNAs. If this can be raised, overall efficiency will be improved, so one potential extension of this work would be a more thorough comparison of mRNA editors and modifications to improve their function.

References:

1 Nissim, L. Cell. vol. 54, 698-710, 2014.

2 Filonov, G.S., Jour. of Amer. Chem. Soc. vol. 146, 16299-16308, 2014.

Enhancer Composition and Dosage Control Developmental Gene Expression

Conference: International Conference on CRISPR Technologies

Year: 2017

Proceeding: International Conference on CRISPR Technologies

Group: General Submissions

Session: Genome editing and gene regulation in human health

Time: Monday, December 4, 2017 - 12:50pm-1:15am

Authors:

Will, A., 1. Max Planck Institute for Molecular Genetics
Cova, G., 1. Max Planck Institute for Molecular Genetics
Osterwalder, M., MS 84-171
Chan, W. L., Charité–Universitätsmedizin Berlin
Wittler, L.
Brieske, N.
Heinrich, V.
De Villartay, J. P.
Vingron, M., Max Planck Institute for Molecular Genetics
Visel, A., Lawrence Berkeley National Laboratory
Klopocki, E.
Lupiáñez, D., Charité–Universitätsmedizin Berlin
Mundlos, S., Charité–Universitätsmedizin Berlin

Copy number variations (CNVs) often include noncoding sequences and putative enhancers, but how these rearrangements induce disease is poorly understood. We investigated CNVs involving the regulatory landscape of IHH (encoding Indian hedgehog), which cause multiple, highly localized phenotypes including craniosynostosis and synpolydactyly. We show through transgenic reporter and genome-editing studies in mice that Ihh is regulated by a constellation of at least nine enhancers with individual tissue specificities in the digit anlagen, growth plates, skull sutures and fingertips. Consecutive deletions, resulting in growth defects of the skull and long bones, showed that these enhancers function in an additive manner. Duplications, in contrast, caused not only dose-dependent upregulation but also misexpression ofIhh, leading to abnormal phalanges, fusion of sutures and syndactyly. Thus, precise spatiotemporal control of developmental gene expression is achieved by complex multipartite enhancer ensembles. Alterations in the composition of such clusters can result in gene misexpression and disease.

Engineered Stem Cell Models to Reconstitute Kidney Function and Disease

Conference: International Conference on CRISPR Technologies

Year: 2017

Proceeding: International Conference on CRISPR Technologies

Group: General Submissions

Session: Genome editing and gene regulation in human health

Time: Monday, December 4, 2017 - 2:30am-2:55am

Authors:

Musah, S., Harvard University
Ingber, D. E., Harvard Medical School & Children's Hospital Boston
Church, G. M., Harvard Medical School

Glomerulopathy is a leading cause of end-stage renal failure. An in vitromodel of human glomerulus could facilitate therapeutic discovery and illuminate kidney disease mechanisms. Efforts to develop such models are limited by the lack of functional human podocytes, the specialized epithelial cells that encase glomerular capillaries and regulate selective permeability in the glomerulus. We developed a highly efficient method for directed differentiation of human induced pluripotent stem (iPS) cells into mature podocytes. By using microfluidic organs-on-chips devices, we engineered an in vitro model of the human glomerular capillary wall that supports podocyte differentiation and recapitulate the normal tissue-tissue interface (podocytes—basement membrane—endothelium) and selective molecular filtration of the glomerulus. The engineered human glomerulus-on-a-chip also replicates kidney disease phenotypes in vitro. Our results demonstrate the feasibility of generating mature podocytes in a robust manner, providing an opportunity to engineer a functional human glomerular capillary wall-on-a-chip. This work could illuminate the mechanisms of human kidney disease, and provide a functional in vitro system for therapeutic development. These results also provide opportunities for personalized medicine, cell therapy, and regenerative medicine.

CRISPR-Based Synthetic Gene Circuits for Controllable Gene Therapies

Conference: International Conference on CRISPR Technologies

Year: 2017

Proceeding: International Conference on CRISPR Technologies

Group: General Submissions

Session: Genome editing and gene regulation in human health

Time: Monday, December 4, 2017 - 1:40am-2:05am

Authors:

Kiani, S., Arizona State University

Clustered regularly interspaced short palindromic repeats (CRISPR) technology has opened-up widespread applications including genetic disease modeling, functional screens, and synthetic gene regulation. The plausibility of in vivo genetic engineering using CRISPR has garnered significant traction as a next generation in vivo therapeutic. However, safety and controllability of CRISPR therapeutics will remain a challenge for human translation. To this end, synthetic biology can offer important solutions. We apply the design principles of synthetic biology to develop CRISPR-based genetic logic gates, layered genetic circuits and safety switches in mammalian cells. Synthetic CRISPR-based genetic circuits can be designed to generate a user-defined actuation such as spatiotemporal regulation. Here, I will review my team's efforts to engineer CRISPR for safer and controllable gene therapies. Engineering gRNA and Pol II-driven control mechanisms enable cell and context specific regulation. This strategy combined with synthetic promoter engineering allow us to develop genetic circuits to control the timing and sequence of function of catalytically active or inactive Cas9 protein or the gRNA. In parallel, we have developed and tested a number of safety switches in mammalian cells by harnessing multi-functionality of CRISPR, including DNA cleavage or transcriptional modulation or both. For some clinical applications, transcriptional modulation by nuclease incompetent Cas9 can be a safer alternative that genetic disruption by Cas9 nuclease. We utilize the nuclease incompetent Cas9 protein to test a plethora of potent transcriptional activation and repression domains, test them both in synthetic genetic circuits and endogenous gene circuits to understand design principles of transcriptional activation and repression by CRISPR and use for therapeutic purposes. We have engineered potent and novel CRISPR repressors that can be used to generate two layered Pol II driven genetic cascades. An ideal in vivo gene therapy platform provides safe, reprogrammable, and precise strategies which modulate cell and tissue gene regulatory networks with a high temporal and spatial resolution. We believe that synthetic CRISPR-based circuits can push the boundaries of CRISPR technologies towards such capacity for future therapies.

Computational Selection of CRISPR/Cas9 Guide-RNAs to Target Highly Variable Integrated HIV-1 Provirus Using Next Generation Sequencing

Conference: International Conference on CRISPR Technologies

Year: 2017

Proceeding: International Conference on CRISPR Technologies

Group: General Submissions

Session: Genome editing and gene regulation in human health

Time: Monday, December 4, 2017 - 2:55am-3:20am

Authors:

Dampier, W., Drexel University College of Medicine
Nonnemacher, M., Drexel University College of Medicine
Chung, C. H., Drexel University College of Medicine
Allen, A., Drexel University College of Medicine, Philadelphia
Sullivan, N., Center for Molecular Virology and Translational Neuroscience
Pirrone, V., Center for Molecular Virology and Translational Neuroscience
Kercher, K., Center for Molecular Virology and Translational Neuroscience
Passic, S., Center for Molecular Virology and Translational Neuroscience
Williams, J., Center for Molecular Virology and Translational Neuroscience
Wigdahl, B., Center for Molecular Virology & Translational Neuroscience

HIV-1 persistence and/or latency even after long-term antiretroviral therapy is a major hurdle to a cure. Genomic editing techniques, like the CRISPR/Cas9 system, hold promise to permanently excise the integrated provirus from a host cell. The CRISPR/Cas9 system could be directed to target the HIV-1 sequence by a guide-RNA designed to be homologous to the virus.integrated provirus. However, targeting is stymied by the rapid accumulation of mutations introduced during HIV-1 replication creating a set of distinct genomic variants that varies within an individual through time and between individuals. Presented here is a computational methodology for designing gRNA sequences to effectively cleave a patient's HIV-1 quasispecies. PBMC genomic DNA was isolated from patients in the Drexel Medicine CNS AIDS Research and Eradication Study (CARES) Cohort and the long terminal repeat (LTR) of the HIV-1 quasispecies was sampled using Next Generation Sequencing (NGS) to a depth of at least 10,000X across 269 patients. gRNAs were computationally selected by examining their predicted binding potential across a random training set of 100 CARES Cohort patient samples. This screening selected a package of 4 or 10 gRNAs, which in silico cleaved the entire detectable quasispecies of the remaining, unseen, 169 CARES samples an average of 3.4 +/- 1.7 or 5.4 +/- 3 times, respectively. The package was further tested against an independent national sampling of North American subtype B LTRs from the Los Alamos National Database and in silico was shown to cleave all LANL LTR sequences. Functional expression of the gRNAs in a P4R5 cell line model have shown knockdown of LTR activity. This work presents a step towards understanding the complex task of using excision therapy to target HIV-1 quasispecies in the infected patient population.

A Novel Cas9 Mutant Confers Reduced Off-Target Gene Editing While Maintaining on-Target Potency

Conference: International Conference on CRISPR Technologies

Year: 2017

Proceeding: International Conference on CRISPR Technologies

Group: General Submissions

Session: Genome editing and gene regulation in human health

Time: Monday, December 4, 2017 - 12:00pm-12:25pm

Authors:

Cannieux, M.
Rettig, G., Integrated DNA Technologies
Thommandru, B.
Jacobi, A., Integrated DNA Technologies
Collingwood, M.
Schubert, M.
Behlke, M.
Vakulskas, C.
Lister, J.

We will present the screening of a Cas9 variant with increased specificity from a library of 250,000 recombinant clones. We will demonstrate robust transfection efficiency by lipofection and electroporation as RiboNucleoProtein (RNP) in various established or primary cell lines.

We will verify that this Cas9 mutant combines strongly reduced off-target editing with maintained on-target activity, compared with the Wild-type S.p Cas9 nuclease, when used as RNP. We will review high efficiency Knock-In applications with both short (<120 nt) single-stranded donor oligonucleotides and long (up to 2kb) single-stranded donor DNA fragments.

One-Step Assembly of Class 2 CRISPR Arrays for Multiplexed Cleavage and Regulation

Conference: International Conference on CRISPR Technologies

Year: 2017

Proceeding: International Conference on CRISPR Technologies

Group: General Submissions

Session: Session 2: Genome editing and gene regulation in industrial bacterial biotechnology

Time: Monday, December 4, 2017 - 12:00pm-12:25pm

Authors:

Liao, C., Helmholtz Institute for RNA-based Infection Research
Beisel, C. L., North Carolina State University
Ttofali, F., North Carolina State University
Slotkowski, R. A., North Carolina State University
Denny, S. R., North Carolina State University
Keung, A., North Carolina State University
Leenay, R. T., North Carolina State University

Clustered Regularly Interspaced short palindromic Repeats (CRISPR) systems have gained widespread attention because they can be harnessed for genome editing and gene regulation in different organisms. Most Class 2 CRISPR effector proteins discovered so far have the ability to process CRISPR arrays into individual guide RNAs. However, the conserved repeat elements in the CRISPR array prevent its chemical synthesis, limiting our ability to use these arrays for multiplexed genome editing and gene regulation. To overcome this barrier, we have developed a novel assembly scheme for the single-step, efficient generation of large CRISPR arrays. The resulting arrays guided multiplexed DNA cleavage and gene regulation in Escherichia coli and Saccharomyces cerevisiae by the FnCpf1 and AsCpf1 nucleases. The cloning scheme was also extended to the generation of CRISPR arrays utilized by Cas9 for multiplexed DNA cleavage in E. coli and by C2c2 for multiplexed targeting in TXTL. Moreover, CRISPR arrays co-utilized by Cpf1 and Cas9 were also assembled and proved to be functional in E. coli. We also noticed that some functional spacers showed reduced or negligible activity in the context of CRISPR arrays. This platform offer simplified multiplexing with Cpf1 and other Class 2 effector proteins across the myriad applications of CRISPR technologies.

Genotyping of Probiotic Bifidobacterium Longum Strains Using CRISPR Loci

Conference: International Conference on CRISPR Technologies

Year: 2017

Proceeding: International Conference on CRISPR Technologies

Group: General Submissions

Session: Session 2: Genome editing and gene regulation in industrial bacterial biotechnology

Time: Monday, December 4, 2017 - 12:25pm-12:50pm

Authors:

Claudio, H. C., NCSU
Barrangou, R., North Carolina State University
Crawley, A., NCSU
Borja, S., Dairy Research Institute of Asturias

Bifidobacterium longum strains are amongst the most widely used probiotics for human consumption due to their ability to modulate the host immune response, and withstand passage through the human gastrointestinal tract. CRISPR-Cas systems occur widely in bacteria and provide adaptive immunity via acquisition of DNA sequences from viruses and plasmids that constitute unique genetic records of immunization. We performed bioinformatics analyses of 66 B. longumgenomes that identified 29 CRISPR loci in 25 strains. Each CRISPR locus was characterized and used for phylogenetic analyses. We identify various Type I and Type II CRISPR-Cas systems that are widespread in this species, notably I-C, I-E and II-C. Noteworthy, Type I-C systems showed extended CRISPR arrays, with extensive spacer diversity. The spacer sequences enable high-resolution genotyping of B. longum strains and revealed a common origin for some strains reflecting divergent evolution under putative selective pressure from invasive DNA. Analysis of CRISPR spacer origin also revealed their implication in phages and prophages targeting and novel PAM sequences. Altogether, results suggest that these immune systems are likely active in this species, and provide a basis for high-resolution genotyping. Furthermore, this study opens new avenues for the repurposing of the endogenous Cas machinery for genome editing applications to enhance their probiotic effect.

Evolutionary Robust Many-Regulator Crispri Arrays for Large-Scale Reprogramming of Metabolism and Cellular Physiology

Conference: International Conference on CRISPR Technologies

Year: 2017

Proceeding: International Conference on CRISPR Technologies

Group: General Submissions

Session: Session 2: Genome editing and gene regulation in industrial bacterial biotechnology

Time: Monday, December 4, 2017 - 1:40am-2:05am

Authors:

Reis, A., Penn State University
Halper, S., Penn State Univ
Hossain, A., Pennsylvania State University
Salis, H. M., Pennsylvania State University
Vezeau, G., Pennsylvania State University
Clauer, P., Pennsylvania State University

We present a new CRISPR approach to simultaneously regulate the expression of up to 20 genes within an endogenous metabolic or signaling network, utilizing a newly developed toolbox of non-repetitive CRISPRi genetic parts designed for maximum evolutionary robustness. CRISPRi employs the deactivated version of Cas9 and single-guide RNA (sgRNA) regulators to bind to targeted DNA sites and repress transcription. Here, we show how to rationally design non-repetitive parts and optimally combine them together into Very Long sgRNA Arrays (VLSA) to implement complex metabolic redirections and cellular phenotypes. We demonstrated this technology by building and characterizing VLSAs for several applications: reprogramming metabolism to over-produce desired chemicals, creating auxotrophic GMO containment strains, and triggering increased cellular susceptibility to therapeutics. Notably, VLSAs can be synthesized as single DNA fragments and integrated into host genomes, efficiently and quickly, greatly accelerating the Synthetic Biology design-build-test cycle.

Development of a Conjugation-Based Genome Editing Tool in Shewanella Oneidensis MR-1 Using CRISPR/Cas9 and Recombineering

Conference: International Conference on CRISPR Technologies

Year: 2017

Proceeding: International Conference on CRISPR Technologies

Group: General Submissions

Session: Session 2: Genome editing and gene regulation in industrial bacterial biotechnology

Time: Monday, December 4, 2017 - 1:15am-1:40am

Authors:

Domènech Corts, A., University of Colorado
Gill, R. T., University of Colorado Boulder
Gralnick, J. A., University of Minnesota

Development of a conjugation-based genome editing tool in Shewanella oneidensis MR-1 using CRISPR/Cas9 and recombineering

Shewanella oneidensis MR-1 is a facultative anaerobic bacterium capable of respiring a vast number of electron acceptors including insoluble toxic metals, radionuclides, and carbon electrodes. The metabolic pathways and mechanisms by which extracellular electron transfer occurs in MR-1, however, are still poorly understood. Considering the potential application of this organism towards bioremediation of contaminated aquatic and subsurface environments as well as microbial fuel cells, it is important to have efficient genome-editing tools that allow us to easily engineer this organism for better overall understanding of its biological capabilities. While MR-1 is amenable to targeted genomic mutagenesis techniques, current technologies are laborious, time-consuming, and must be conducted individually rather than collectively.

Recently, a technology utilizing a high-throughput approach coupling massive oligomer synthesis to CRISPR/Cas9 and λ-Red systems, named CREATE (CRISPR-enabled trackable genome engineering) has been demonstrated to be a powerful tool for highly efficient and precise targeted genomic editing in Escherichia coli. Introduction of CREATE intoE. coli, however, requires the use of electroporation to deliver the system into the cell. While Shewanella and other environmentally relevant microbes are recalcitrant to electroporation, conjugation is a commonly used technique allowing for plasmid transfer and uptake.

Here, we show our progress towards the development of a conjugation-based CREATE system for use in MR-1. Conjugation-based CREATE will utilize the Streptococcus pyogenes type II CRISPR/Cas9 system as well as phage recombineering as previously demonstrated in E. coli. We have optimized existing MR-1 conjugation protocols for the purpose of introducing a CREATE library and provide evidence of efficient CRISPR/Cas9 function in MR-1 as a selection. Additionally, we are developing a minimal inducible system consisting of Cas9 and recombineering proteins in one plasmid for ease portability and adaptability across microbial species.

Our work primarily aims to expand the metabolic engineering toolbox of the environmentally relevant organism MR-1 allowing for expansion of bioremediation technologies. Additionally, this developed conjugation-based CREATE system can be broadly applied to other microbes that are intractable via electroporation and non-recombinogenic, while simultaneously overcoming the dependence upon suicide vectors and/or random mutagenesis approaches.

CRISPR-Cas Activation (CRISPRa) for Transcriptional Control in E. coli

Conference: International Conference on CRISPR Technologies

Year: 2017

Proceeding: International Conference on CRISPR Technologies

Group: General Submissions

Session: Session 2: Genome editing and gene regulation in industrial bacterial biotechnology

Time: Monday, December 4, 2017 - 12:50pm-1:15am

Authors:

Fontana, J.
Dong, C., University of Washington
Ham, J.
Patel, A.
Stevens, J., University of Washington
Carothers, J., University of Washington
Zalatan, J.

Metabolic engineering efforts often require controlling the expression level of multiple genes. We have developed a new suite of synthetic bacterial transcriptional activators by linking activation domains to the CRISPR-Cas system. Together with existing tools for CRISPRi gene repression, these synthetic activators enable programmable transcriptional up- and down-regulation over multiple endogenous and heterologous targets at the same time. Interestingly, effective gene activation requires target sites situated in a narrow region just upstream of the gene of interest, in sharp contrast to the relatively flexible target site requirements for gene activation in eukaryotic cells. We present a strategy for synthetic promoter design to independently activate multiple heterologous genes via CRISPRa. Last, we demonstrate the bioproduction of industrially-relevant aromatics (p-aminocinnamic acid) via CRISPRa control of heterologous gene expression in E. coli. In this talk, I will present these results and describe on-going work to fine-tune CRISPRa transcriptional activation, engineer dynamically-controlled CRISPRa systems and build more complex synthetic bacterial devices with a wide range of applications in industrial biosynthesis.

CRISPR Hinderance of Adaptation of Organisms (CHAOS) Based Therapeutics

Conference: International Conference on CRISPR Technologies

Year: 2017

Proceeding: International Conference on CRISPR Technologies

Group: General Submissions

Session: Session 2: Genome editing and gene regulation in industrial bacterial biotechnology

Time: Monday, December 4, 2017 - 12:00pm-12:25pm

Authors:

Chatterjee, A., University of Colorado Boulder

The emergence of multidrug-resistant "superbugs" continues to pose a looming global health crisis that necessitates novel therapeutic strategies to curb their threat. The advent of CRISPR technologies has expanded our capacity to investigate such strategies. In particular, deactivated versions of the Cas9 (dCas9) enzyme enable relatively facile manipulation of gene expression in a multiplexed fashion that was once prohibitively difficult and expensive to perform. Employing dCas9 with multiplexed sgRNAs permits manipulation of specific bacterial gene expression profiles during the critical stage of early antibiotic exposure. Previous work has consistently revealed that upon exposure to antibiotics, bacterial transcriptomes adjust to new levels that impart improved fitness phenotypes. This arises from the natural variation in gene expression across the population that exists as a bet-hedging tool to enable subsets of the population to survive during sudden environmental changes. We postulated that this heterogeneity can act as a "double-edged sword" – if a subset of random expression states provides an advantage to stress, then another subset will similarly prove disadvantageous. Whether or not a particular expression state is advantageous or deleterious depends upon the degree of epistasis between each change. Epistasis describes the nonlinear interactions between two or more simultaneous genetic changes and is widely recognized for its role in shaping evolutionary trajectories. While research into epistasis has largely focused on simultaneous mutations, similar effects have been observed from changes in gene expression. In this study, we manipulated gene expression epistasis using multiplexed dCas9 perturbations in an attempt to curb bacterial adaption to antibiotics. Here we present evidence that multiplexed dCas9 perturbations induce significant negative epistasis, and reduce fitness of *Escherichia coli* more than 50% during antibiotic exposure. Greater epistasis correlates with increasing nodes within affected protein interaction networks, with perturbations affecting essential metabolic pathways appearing to induce the strongest epistasis. Furthermore, greater negative epistasis corresponds with significantly diminished adaptation rates, thereby increasing antibiotic susceptibility over time. This suggests that multiplexed dCas9 perturbations can restrict adaptation and curb the emergence of antibiotic resistance. Such phenomenon could curtail antibiotic resistance evolution when administered as a co-therapy with last-resort antibiotics.

Characterizing a Thermostable Cas9 and Exploring Its Potential As a Genome Engineering Tool for Both Thermophilic and Mesophilic Prokaryotes

Conference: International Conference on CRISPR Technologies

Year: 2017

Proceeding: International Conference on CRISPR Technologies

Group: General Submissions

Session: Session 2: Genome editing and gene regulation in industrial bacterial biotechnology

Time: Monday, December 4, 2017 - 3:20am-3:45am

Authors:

Mougiakos, I., Wageningen UR
Mohanraju, P., Wageningen University
Bosma, E. F., DTU Biosustain
Finger Bou, M., Wageningen University
Vrouwe, V., Wageningen University and Research
Naduthodi, M., Wageningen University and Research
Gussak, A., Wageningen University and Research
Brinkman, R., Corbion
van Kranenburg, R., Wageningen University
van der Oost, J., Wageningen University

CRISPR-Cas9 based genome engineering tools have revolutionized fundamental research and biotechnological exploitation of both eukaryotes and prokaryotes. However, the mesophilic nature of the already established Cas9 systems restricts their applicability, excluding thermophiles and cellular environments in which mesophilic enzymes are easily inactivated. Here, we identify and characterize Cas9 from the CRISPR-Cas type-IIC system of the thermophilic bacterium *Geobacillus thermodenitrificans* T12 as an RNA-guided DNA endonuclease. We provide fundamental insights

into a thermophilic CRISPR-Cas family member and describe our work towards the development of

a genome editing and silencing tool readily applicable to thermophiles.

Casper: Algorithms and Integrated Software Platform for Precise CRISPR Genome Editing of Single and Consortia of Organisms

Conference: International Conference on CRISPR Technologies

Year: 2017

Proceeding: International Conference on CRISPR Technologies

Group: General Submissions

Session: Session 2: Genome editing and gene regulation in industrial bacterial biotechnology

Time: Monday, December 4, 2017 - 12:00pm-12:25pm

Authors:

Trinh, C. T., University of Tennessee
Mendoza, B., University of Tennessee Knoxville

Genetic diversity of non-model organisms offers a repertoire of unique phenotypic features for exploration and cultivation in synthetic biology and metabolic engineering applications. To realize this enormous potential, it is critical to have an efficient genome editing tool for rapid strain engineering of these organisms to perform novel programmed functions. To accommodate the use of CRISPR/Cas systems for genome editing across organisms, we have developed algorithms and associated software, named CASPER (CRISPR Associated Software for Pathway Engineering and Research), for identifying on- and off-targets with enhanced predictability coupled with an analysis of non-unique (repeated) targets to assist in editing any organism with various endonucleases. Utilizing CASPER, we demonstrated a modest 2.4% and significant 30.2% improvement (F-test, $p<0.05$) over the conventional methods for predicting on- and off-target activities, respectively. Further we used CASPER to develop novel applications in genome editing: multitargeting analysis (i.e. simultaneous multiple-site modification on a target genome with a sole guide-RNA (gRNA) requirement) and multispecies population analysis (i.e. gRNA design for genome editing across a consortium of organisms). Our analysis on a selection of industrially relevant organisms revealed a number of non-unique target sites associated with genes and transposable elements that can be used as potential sites for multitargeting. The analysis also identified shared and unshared targets that enable genome editing of single or multiple genomes in a consortium of interest. Utilizing multiple algorithms for on- and off-target analysis of the guided RNA/Cas complex, CASPER achieves higher accuracy at predicting experimental results. Further, CASPER enables end-users to design guided RNAs to target multiple chromosome loci at once and across a consortium of organisms. We envision CASPER as a useful platform to enhance the precise CRISPR genome editing for metabolic engineering and synthetic biology applications.

A Thermophilic Cas9 Specific for Tandem Cytosine-Adjacent DNA and Sensitive to DNA Topology

Conference: International Conference on CRISPR Technologies

Year: 2017

Proceeding: International Conference on CRISPR Technologies

Group: General Submissions

Session: Session 2: Genome editing and gene regulation in industrial bacterial biotechnology

Time: Monday, December 4, 2017 - 2:05am-2:30am

Authors:

Li, H.

Cas9 is an RNA-guided DNA cleavage enzyme being actively developed for genome editing and gene regulation. To be cleaved by Cas9, a double stranded DNA, or the protospacer, must be complementary to the Cas9-bound guide RNA and adjacent to a short Cas9-specific element called Protospacer Adjacent Motif (PAM). Understanding the correct juxtaposition in time and space of the protospacer- and PAM-interaction with Cas9 will enable development of versatile and safe Cas9-based technology. We report identification and biochemical characterization of Cas9 from thermophile Acidothermus cellulolyticus (AceCas9).Acidothermus cellulolyticus has been used for producing the enzyme endo-1, 4-β-glucanase (E1) which is used for the commercial hydrolysis of cellulose into glucose. Genome editing in A. cellulylyticus could impact its utilization in bioenergy development. We found that AceCas9 depends strictly on a 5'-NNNCC-3' PAM and is more efficient in cleaving negative supercoils than relaxed DNA. We further characterized the dependence of AceCas9 on temperature, guide length, divalent metal ions, and mismatches to the guide RNA. The thermostability, cytosine-specific and DNA topology-sensitive properties of the AceCas9 maybe explored for specific genome editing applications.

Transcriptional Reprogramming in Yeast Using dCas9 and Combinatorial gRNA Strategies

Conference: International Conference on CRISPR Technologies

Year: 2017

Proceeding: International Conference on CRISPR Technologies

Group: General Submissions

Session: Session 3: Genome editing and gene regulation in industrial eukaryotic biotechnology

Time: Monday, December 4, 2017 - 1:15am-1:40am

Authors:

Damgaard Jensen, E., Technical University of Denmark
Jakociunas, T., The Novo Nordisk Foundation Center for Biosustainability
Jensen, M. K., Technical University of Denmark
Keasling, J. D., Technical University of Denmark

Transcriptional regulation is a fundamental and vital process occurring in living organisms. Synthetic biology tools for orthogonal control of transcription have been addressed in a large number of studies to harness this virtue of life for increased metabolic flux and product yield. Extending from previous efforts, the nuclease-deficient Cas9 (dCas9) protrudes as a recent and flexible tool towards genome reprogramming. We approached dCas9-mediated transcriptional reprogramming inSaccharomyces cerevisiae using scaffold-RNAs to recruit transcriptional effectors [1] that were subsequently targeted to genes controlling production of isoprenoids in yeast. Our investigations included >100 guide-RNAs (gRNAs) on 14 different yeast promoters. Through multiplex reprogramming we were able to perturb fluxes in targeted pathways based on phenotypic screening and quantitative measurements.

However, the study also highlighted issues in using dCas9 for transcriptional reprogramming in yeast; most of our gRNA library did not significantly de- or increase expression levels in our reporter assay, where GFP was directly fused to each promoter under investigation. We tested multiple gRNAs per promoter and investigated different features such as protospacer adjacent motif distances to transcriptional start sites.

Taken together, this study presented multiplex designs that can significantly perturb yeast isoprenoid production by using dCas9 and scaffold-RNAs for transcriptional reprogramming. This study also contributed perspective to transcriptional reprogramming using dCas9 in baker's yeast, and our findings indicate that basic understanding of underlying mechanisms require further exploration, before genome reprogramming can be fully predictive and applied in yeast.

Reference

1. Zalatan JG, Lee ME, Almeida R, Gilbert LA, Whitehead EH, La Russa M, et al. Engineering Complex Synthetic Transcriptional Programs with CRISPR RNA Scaffolds. Cell. 2014;160:339–50. doi:10.1016/j.cell.2014.11.052.

Streamlining and Enhancing Guide RNA Expression for Programmable Large-Scale Multiplex of Desired Genes in Saccharomyces Cerevisiae

Conference: International Conference on CRISPR Technologies

Year: 2017

Proceeding: International Conference on CRISPR Technologies

Group: General Submissions

Session: Session 3: Genome editing and gene regulation in industrial eukaryotic biotechnology

Time: Monday, December 4, 2017 - 12:25pm-12:50pm

Authors:

Deaner, M. H., The University of Texas at Austin
Alper, H., The University of Texas at Austin

While metabolic engineering has successfully led to the renewable production of high-value chemicals such as artemisinin and 1,3-propanediol, it nonetheless suffers from the slow rate of genomic modification. As a solution, the CRISPR-dCas9 system (typically with fused activator or repressor domains) has been used to enable modification of gene expression in a rapid, transferrable manner. This talk covers three major advances we have made in adapting and expanding this system for the yeast Saccharomyces cerevisiae.

First, to bypass the typical binary on/off gene modulation inherent in this system, we previously developed a scheme for graded dCas9 regulation with a large dynamic range of perturbation. We accomplish this by modulating dCas9-activator/repressor proximity relative to the desired gene target. Linking dCas9-mediated graded gene expression with desired phenotypes to Systematically Test Enzyme Perturbation Sensitivities (STEPS) afforded rapid iterative optimization of glycerol and 3-dehydroshikimate titers (5.7 and 7.8-fold improvement, respectively). While this approach was a step forward for more rapid strain engineering, it can solely access a small combinatorial expression space due to the use of repetitive and bulky Pol III expression cassettes for each individual sgRNA.

Second, we thus address the combinatorial optimization challenge by developing a streamlined approach for gene modulation. This approach uses Pol II promoters and ribozymes to express multiple sgRNAs from a single cistron while at the same time re-purposing the dCas9-VPR activator to predictably activate or repress target genes based on its targeting location within the desired gene locus. In doing so, we achieve an unexpected yet beneficial result-that strong Pol II expression of sgRNAs produces nearly 4-fold more mature sgRNA than the traditional Pol III SNR52 promoter, thus affording stronger CRISPRi repression and doubling the dynamic range of dCas9-VPR modulation of the strong TPI1promoter.

Third, we demonstrate a final streamline of our approach that optimizes the expression cassette for more efficient excision of sgRNAs via endogenous RNA processing. Ultimately, these more optimized elements further enhance dCas9 perturbation efficiency, reduce the DNA load, and facilitate easy cloning schemes for rapid combinatorial assembly of arrays that can modify entire pathways. Using this system, we demonstrate that we can rapidly synthesize and transform a desired, complex genotype into cells and showcase the portability of this approach across multiple different strains. We foresee this approach as a tool to rapidly condense genome regulation into a singular cassette that can be transferred to different strain backgrounds by a singular transformation step.

Gene Excision By Dual-Cutting CRISPR-Cas9 in Yarrowia Lipolytica

Conference: International Conference on CRISPR Technologies

Year: 2017

Proceeding: International Conference on CRISPR Technologies

Group: General Submissions

Session: Session 3: Genome editing and gene regulation in industrial eukaryotic biotechnology

Time: Monday, December 4, 2017 - 2:30am-2:55am

Authors:

Blenner, M., Clemson University

A genome editing tool has become increasingly required for metabolic engineering of non-model and non-conventional microorganism. For example, a CRISPR-Cas9 tool for the oleaginous yeast, Yarrowia lipolytica, has been used for promoting double-strand breaks. When provided with a homologous template, Cas9 activity at certain sites can lead to homologous recombination. In the absence of a homologous donor, indel mutations can occur leading to gene inactivation. However, as mutations are often the insertion or deletion of a single to three nucleotides, the activity of the edited gene may remain unaltered. Furthermore, the remaining unedited portions of targeted genes have proved problematic due to gene conversion events. Here, we describe the development and characterization of a gene excision tool for Yarrowia lipolytica. This tool has been demonstrated on five targets and has successfully removed segments as large as 4kb. The core principle underlying the gene excision methodology is the simultaneous double-stand break at two locations, and non-homologous end joining repair. Given that both breaks need to occur during the same cell division, and that these breaks have been repaired together, a 12-16% efficiency was greater than expected. Since the resulting genomic region has a significant size difference, screening by colony PCR is simple compared with alternative methods for identifying indels. Recent efforts to multiplex and improve the overall efficiency of this technology will be presented.

CRISPR-Cas9 Enabled Metabolic Engineering of Oleaginous and Thermotolerant Yeasts

Conference: International Conference on CRISPR Technologies

Year: 2017

Proceeding: International Conference on CRISPR Technologies

Group: General Submissions

Session: Session 3: Genome editing and gene regulation in industrial eukaryotic biotechnology

Time: Monday, December 4, 2017 - 2:55am-3:20am

Authors:

Schwartz, C., UC Riverside
Löbs, A. K., UC Riverside
Wheeldon, I., UC Riverside

Identification of organisms which natively possess desirable phenotypes and application of advanced engineering techniques to enhance and understand these properties is a valuable approach for strain engineering in synthetic biology and metabolic engineering. Genome engineering in these non-model organisms has been significantly advanced by the widespread adoption of the type II CRISPR-Cas9 system. We have adapted CRISPR-Cas9 from Streptococcus pyogenes for genome editing in the yeasts Yarrowia lipolytica, of interest due to its ability to synthesize and accumulate lipids, and Kluyveromyces marxianus, which has a native capacity to synthesize volatile esters and ethanol at high temperatures. To overcome low gene disruption efficiencies, we designed synthetic RNA polymerase III promoters for guide RNA expression. These hybrid promoters enhanced CRISPR-Cas9 function in both organisms. In Y. lipolytica, the CRISPR-Cas9 system was applied to markerlessly integrate genes into well-characterized genomic loci for standardized pathway engineering, yielding strains that synthesize the carotenoid lycopene. We further applied the developed system for gene expression regulation via CRISPR interference (CRISPRi), which allowed us to enhance homologous recombination by transiently repressing genes involved in nonhomologous end-joining. In K. marxianus, a knockout library of alcohol dehydrogenases was generated to enable fundamental studies of gene function and allowed the identification of novel ethyl acetate synthesis activity. A CRISPRi system was then developed and used to shift metabolism away from respiration and towards fermentation to improve ester production. Together, these applications show the potential of CRISPR-Cas9 to enable advanced genome editing and transcriptional regulation in organisms with favorable native phenotypes

Using Nanoprecipitation to Design and Engineer PLGA Nanoparticle Carriers Loaded CRISPR-Cas9 Plasmids for Increased Delivery and Transfection

Conference: International Conference on CRISPR Technologies

Year: 2017

Proceeding: International Conference on CRISPR Technologies

Group: General Submissions

Session: Achieving efficient delivery and editing

Time: Tuesday, December 5, 2017 - 1:15am-1:40am

Authors:

Jo, A., Macromolecules Innovation Institute
Ringel-Scaia, V., Virginia Tech
Allen, I., Macromolecules Innovation Institute
Davis, R. M., Virginia Tech

One of the principal problems in implementing CRISPR-Cas9 in gene therapy applications is the difficulty in effective delivery of high molecular weight plasmid DNA into cells in vivo. Previous gene transfection approaches have used both chemical and physical stimuli, but these often destabilize cellular membranes resulting in unacceptable cytotoxicity.[1] Polymer nanoparticles can protect the plasmid from degradation en route and increase delivery into the cell without the harmful side effects seen using more traditional methods for transfection.

Using an FDA-approved polymer and a modified nanoprecipitation method[2], a poly(lactic-co-glycolic acid) (PLGA) nanoparticle carrier was engineered to deliver a CRISPR-Cas9 plasmid designed to induce a single point mutation in the mouse Tlr4 gene with the intention of inducing a loss of immune response to stimuli, such as lipopolysaccharides (LPS). The PLGA carrier was also fluorescently labeled by encapsulation of a commercially available fluorophore, 6,13-Bis(triisopropylsilylethynl) pentacene (TIPS pentacene), previously used in similar polymeric nanoparticles and shown to effectively visualize successful uptake of particles into macrophages.[3] This previous work suggests that the PLGA particles will also be readily phagocytosed by macrophages and be an excellent candidate to evaluate the release of payload plasmid within the cell.

The main focus of this presentation will be on the particle processing and engineering that preceded the cell and animal studies. This work addresses the questions of solvent system, size and surface chemistry, carrier composition and DNA loading, as well as cell uptake and transfection efficacy. Initial studies demonstrate a loading of 0.61 wt% of the ~8.5 kb CRISPR-Cas9 plasmid using an amine end capped PLGA at an encapsulation efficiency of 82%. These results are already comparable to the typical 1 wt% loadings seen for DNA with only 4.5 kb.[2] Following this promising framework, the next steps include increasing the target DNA loading, and studying the release kinetics of the biologic. This systematic design and characterization of the particles can be translated to a variety of different applications in the area of nanoparticle drug delivery and CRISPR-Cas9 gene therapy.

References

1. Bauer, M.; Kristensen, B. W.; Meyer, M.; Gasser, T.; Widmer, H. R.; Zimmer, J.; Ueffing, M., Toxic effects of lipid-mediated gene transfer in ventral mesencephalic explant cultures. Basic Clin Pharmacol 2006, 98(4), 395-400.

2. Niu, X. M.; Zou, W. W.; Liu, C. X.; Zhang, N.; Fu, C. H., Modified nanoprecipitation method to fabricate DNA-loaded PLGA nanoparticles.Drug Dev Ind Pharm 2009, 35 (11), 1375-1383.

3. McDaniel, D. K.; Jo, A.; Ringel-Scaia, V. M.; Coutermarsh-Ott, S.; Rothschild, D. E.; Powell, M. D.; Zhang, R.; Long, T. E.; Oestreich, K. J.; Riffle, J. S.; Davis, R. M.; Allen, I. C., TIPS pentacene loaded PEO-PDLLA core-shell nanoparticles have similar cellular uptake dynamics in M1 and M2 macrophages and in corresponding in vivo microenvironments.Nanomedicine: Nanotechnology, Biology and Medicine 2017, 13 (3), 1255-1266.

Therapeutic Strategies Via CRISPR-Cas: New Approaches and New Challenges

Conference: International Conference on CRISPR Technologies

Year: 2017

Proceeding: International Conference on CRISPR Technologies

Group: General Submissions

Session: Achieving efficient delivery and editing

Time: Tuesday, December 5, 2017 - 12:00pm-12:25pm

Authors:

Mali, P., University of California San Diego

The recent advent of RNA-guided effectors derived from clustered regularly interspaced short palindromic repeats (CRISPR)–CRISPR-associated (Cas) systems have dramatically transformed our ability to engineer the genomes of diverse organisms. As unique factors capable of co-localizing RNA, DNA, and protein, tools and techniques based on these are paving the way for unprecedented control over cellular organization, regulation, and behavior. Here I will describe some of our ongoing efforts towards engineering this system for enabling therapeutic applications.

Synthetic Sgrna Enables Highly Efficient and Consistent CRISPR Editing of Cells for Automation and Therapeutic Applications

Conference: International Conference on CRISPR Technologies

Year: 2017

Proceeding: International Conference on CRISPR Technologies

Group: General Submissions

Session: Achieving efficient delivery and editing

Time: Tuesday, December 5, 2017 - 12:00pm-12:25pm

Authors:

Holden, K., Synthego

Achieving consistent and high editing efficiencies with CRISPR is critical for automation and therapeutic applications with primary cells, and remains a significant challenge. Through a collaborative effort, we demonstrate that use of synthetic sgRNA for CRISPR yields improved and consistent editing efficiencies that are required for such applications.

Genome editing in vivo with Cas9 RNP bound to gold nanoparticles

Conference: International Conference on CRISPR Technologies

Year: 2017

Proceeding: International Conference on CRISPR Technologies

Group: General Submissions

Session: Achieving efficient delivery and editing

Time: Tuesday, December 5, 2017 - 12:50pm-1:15am

Authors:

Justad Roise, J.
Murthy, N., U.C. Berkeley
Sadlowski, C., UC Berkeley

The synthetic modification of proteins is a powerful method for enhancing their cellular uptake, reducing immunogenicity, and targeting their delivery to diseased tissues. However, covalent modification of proteins, particularly Cas9, frequently leads to protein inactivation and there is great interest in the development of degradable linkages that can be used as a traceless modification to retain protein activity. We report the design and synthesis of a bi-functional self-immolative linker utilizing a disulfide bond, which is chemically inert during synthetic transformations, and undergoes intracellular sulfhydryl-assisted cleavage to release the unmodified protein. The linker was composed of a PEG chain to provide water solubility, a 1,7-eliminating disulfide to prevent protein degradation, and a terminal orthopyridyl discovered to enhance intracellular uptake. This work demonstrates that thiol cleavable self-immolative linkers are a new traceless modification strategy to effectively deliver intact protein, and provides a chemical handle for additional cargo installation. Studies to determine the effects on enhanced editing efficiency and conjugation to endosomal disrupting conjugates are underway. Other Cas9 delivery strategies developed by our group, including the use of nanoparticles for both *in vitro* and *in vivo* delivery, will also be discussed.

DNA Double-Strand Break in an Actively Transcribed Gene Is Accurately Repaired By Transcript-RNA in Cis Via a Homologous Recombination Mechanism

Conference: International Conference on CRISPR Technologies

Year: 2017

Proceeding: International Conference on CRISPR Technologies

Group: General Submissions

Session: Achieving efficient delivery and editing

Time: Tuesday, December 5, 2017 - 1:40am-2:05am

Authors:

Meers, C., Georgia Institute of Technology
Storici, F., Georgia Institute of Technology

Cells are consistently bombarded from a variety of endogenous and exogenous DNA damaging sources, threatening genome stability. To maintain genome integrity, cells have devised pathways to respond to DNA damage. One of the most threatening forms of DNA damage is a double-strand break (DSB). Three major DSB repair mechanisms are known: non-homologous end joining (NHEJ), in which the broken ends of DNA are ligated together, often resulting in short deletions or insertions, microhomology-mediated end joining (MMEJ), which exploits microhomologous sequences to align the broken ends before joining, resulting in deletions flanking the DSB, and homologous recombination (HR), which has higher fidelity of repair requiring an intact donor DNA molecule with extensive homology to the broken DNA ends. With the advent of CRISPR RNA-guided DNA endonuclease enzymes, there is marked interest in understanding these pathways to facilitate accurate genome engineering events.

Recently, we showed that RNA can serve as a template for DSB repair indirectly, in the form of complementary DNA (cDNA), or directly, in the form of transcript-RNA in budding yeast (Keskin et al., Nature 2014; Meers et al., DNA Repair 2016). To better understand the mechanism of RNA-templated DSB repair, we constructed a yeast system in which an HO endonuclease generated DSB is induced in a constitutively transcribed marker DNA gene controlled by the strong promoter pTEF, similar to a break in an actively transcribed gene. In this constitutive system, direct repair by RNA is remarkably dominant compared to repair by cDNA. The frequency of RNA-templated DSB repair in this constitutive system is strongly reduced in the absence of the HR geneRAD52. Interestingly, deletion of genes associated with NHEJ (KU70, LIG4, or POL4) increases the frequency of DSB repair by transcript-RNA with 7-10% of cells surviving the DSB in a ku70 background being repaired by RNA. RNA-templated repair is stimulated by overexpression of reverse transcriptase. Furthermore, we find that functional mismatch repair (MMR) mechanism is required for efficient RNA-templated DSB repair, possibly because it prevents MMEJ. Overall, these results support an HR type of repair mechanism with donor transcript-RNA at a DSB site in constitutively expressed DNA.

We acknowledge support from the GAANN fellowship to C.M P200A150130-16, NIH grant GM115927, NSF grant MCB-1615335, and the HHMI Faculty Scholar grant #55108574 to F.S.

Design Principles for CRISPR Transcriptional Activation

Conference: International Conference on CRISPR Technologies

Year: 2017

Proceeding: International Conference on CRISPR Technologies

Group: General Submissions

Session: Achieving efficient delivery and editing

Time: Tuesday, December 5, 2017 - 2:05am-2:30am

Authors:

LeGraw, R.
Moghadam, F., Arizona State University
Kiani, S., Arizona State University
Ebrahimkhani, M., Mayo Clinic College of Medicine and Science

Clustered regularly interspaced palindromic repeats (CRISPR) technology has advanced rapidly in recent years, but this explosion of interest has generated a multitude of methodologies and left some mechanistic underpinnings of the system unexplored. One such example is CRISPR-based transcriptional activation. In order to systematically investigate this technique we analyzed several components of the system including identity of target gene, location of sgRNA target sites, quantity of sgRNAs delivered, mass of dCas9/gRNA plasmid transfected, transfection vs. transduction, temporal kinetics of activation, and the Synergistic Activation Mediator (SAM) versus the VP64-p65-Rta (VPR) activation domain.

Using an sgRNA-aptamer binding system, we transiently transfected human embryonic kidney (HEK) cells with plasmids expressing the sgRNA, deactivated Cas9 (dCas9), and the aptamer-binding activation domain. We transfected multiple guides in varying combinations targeted to activate Cyp3a4, albumin, OSM, ZGPAT, Fah, and ATF5. High (200ng) and low (25ng) levels of sgRNA and dCas9 were tested in addition to both the SAM and VPR activation domains along with variants of the Cas9 protein (M2, M4, P300 fusion). The expression levels of albumin and Fah following activation were tracked for seven days post-transfection.

Increased expression was achieved for all genes tested, but certain genes could be induced much more than others, perhaps due to varying endogenous expression; albumin expression was increased by more than 2400-fold, whereas Fah was increased approximately 10-fold. Factors found to improve the expression levels included sgRNA target sites on distinct DNA strands, delivery of low levels of Cas9 and sgRNA, and use of the SAM activation system. Interestingly, in investigation of temporal changes in expression, target transcript levels were found to spike at four days post-transfection. Identifying the effects of each of these variables provides a more robust framework to approach experimental design and optimize standard operating procedures.

Genetic Pest Management Technologies to Control Invasive Rodents

Conference: International Conference on CRISPR Technologies

Year: 2017

Proceeding: International Conference on CRISPR Technologies

Group: General Submissions

Session: CRISPR technologies beyond genome editing and gene regulation

Time: Tuesday, December 5, 2017 - 2:05am-2:30am

Authors:

Kanavy, D.
Threadgill, D.

Many strategies exist to manage invasive pests on islands, ranging from poison to trapping, with varying degrees of success. Genetic technologies are increasingly being applied to insect pests, but so far, not to vertebrates. We are implementing a genetic strategy to eradicate invasive mouse populations as another tool for pest control.

Mus musculus, the common house mouse, is one of the most widespread invasive species. Mice threaten human health, agriculture, and biodiversity on many islands, particularly seabirds. Seabirds are endangered indirectly through competition of resources or predators being attracted by the mice or directly with mice attacking chicks and eggs. Rodenticides are the most common method of eradicating mice, but their use leads to poisoning of non-target species and has limited efficacy against mice. An approach that could eliminate non-target species impact would be to engineer daughterless mice linked to a gene drive system for self-sustained propagation. For this project, we have investigated exploiting a naturally occurring gene drive, the t-complex. Using the t^{w2} haplotype of the t-complex, we observed the t^{w2} haplotype being transmitted to offspring with a transmission distortion ratio of 95.3%.

The daughterless phenotype is being accomplished by inserting the Srygene (male sex-determining gene) into an autosome containing the t^{w2} haplotype via CRISPR/Cas9 gene editing. The presence of Sry will induce testis formation, regardless of the sex chromosomes naturally inherited. When Sry is inserted into the t-complex, the desired gene will spread through the population, eliminating female offspring. This model system will support studies to evaluate the effectiveness of crashing an invasive population without adversely affecting other species. While still in the beginning stages, this is a novel idea and once this method has been perfected, it will open the way to use this genetic strategy for the eradication of other invasive mammal species.

Engineering Safe and Efficient Genome Editing Technologies

Conference: International Conference on CRISPR Technologies

Year: 2017

Proceeding: International Conference on CRISPR Technologies

Group: General Submissions

Session: CRISPR technologies beyond genome editing and gene regulation

Time: Tuesday, December 5, 2017 - 2:05am-2:30am

Authors:

Wegrzyn, R., DARPA

Genome editing technologies provide the unprecedented ability to modify genetic material in a manner that is targeted, rapid, adaptable, and broadly accessible. While the potential for societal benefit from these technologies is immense, longer-term ramifications, such as the potential for these tools to impact large populations of organisms and ecosystems over many generations, must also be considered. Traditional biosafety and biosecurity measures including physical biocontainment, research moratoria, self-governance, and regulation are not designed for technologies that are, in fact, explicitly intended for environmental release and are widely available to users who operate outside of conventional institutions. DARPA created the Safe Genes program to gain a fundamental understanding of how gene editing technologies function; devise means to safely, responsibly, and predictably harness them for beneficial ends; and address potential health and security concerns related to their accidental or intentional misuse. The goal of Safe Genes is to build in biosafety for new biotechnologies at their inception, provide a range of options to respond to synthetic genetic threats, and create an understanding of what is possible, probable, and vulnerable with regard to emergent gene editing technologies. Current and future research priorities supporting safe and secure genome editing technologies will be discussed.

Engineered Broad Host Range Phage for Delivery of CRISPR Antimicrobials

Conference: International Conference on CRISPR Technologies

Year: 2017

Proceeding: International Conference on CRISPR Technologies

Group: General Submissions

Session: CRISPR technologies beyond genome editing and gene regulation

Time: Tuesday, December 5, 2017 - 1:15am-1:40am

Authors:

Collias, D., North Carolina State University
Fagen, J., North Carolina State University
Beisel, C. L., North Carolina State University

DNA cleavage by RNA-guided CRISPR nucleases has been shown to be lethal in bacteria [1,2,3] and has given rise to the development of CRISPR antimicrobials. This technology is programmable by modifying the guide to target specific strains of bacteria [1,2,3]. This comes at an opportune time as antibiotic resistance in bacteria continues to increase. Since implementing CRISPR-Cas systems towards antimicrobial applications, the next challenge was broadly delivering these systems in diverse bacterial communities. We developed an engineered, broad host range P1 bacteriophage as a CRISPR antimicrobial delivery vehicle towards various strains of gammaproteobacteria. We were interested in determining if P1 would be able to package its genome while also encoding for a CRISPR nuclease and guide. As a result, the delivery efficiency based on packaged DNA lengths was determined and we were able to confirm that P1 can package and deliver its genome and DNA encoding for a CRISPR nuclease. Furthermore, we identified locations on the P1 genome to incorporate DNA without impacting particle production and subsequent infection into hosts of interest. Finally, we compared production and infection efficiencies of P1 encoding for Cas9 and Cas12a as a head-to-head comparison of single effectors. CRISPR-Cas systems have been demonstrated as a highly specific and programmable antimicrobial technology that is currently limited by the ability to engineer a broad delivery vehicle.

References:

1. Bikard D, Euler CW, Jiang W, Nussenzweig PM, Goldberg GW, Duportet X, Fischetti VA, Marraffini LA: Exploiting CRISPR-Cas nucleases to produce sequence-specific antimicrobials. Nat. Biotechnol. 2014, 32:1146–1150.

2. Citorik RJ, Mimee M, Lu TK: Sequence-specific antimicrobials using efficiently delivered RNA-guided nucleases. Nat. Biotechnol. 2014, 32:1141–1145.

3. Gomaa AA, Klumpe HE, Luo ML, Selle K, Barrangou R, Beisel CL: Programmable removal of bacterial strains by use of genome-targeting CRISPR-Cas systems. MBio 2014, 5:e00928–13.

Employing CRISPR to Identify and Engineer Synergistic Interactions between Antibiotics and Sequence-Specific Treatments

Conference: International Conference on CRISPR Technologies

Year: 2017

Proceeding: International Conference on CRISPR Technologies

Group: General Submissions

Session: CRISPR technologies beyond genome editing and gene regulation

Time: Tuesday, December 5, 2017 - 12:50pm-1:15am

Authors:

Otoupal, P., University of Colorado Boulder
Erickson, K., University of Colorado Boulder
Chatterjee, A., University of Colorado Boulder

Due to the rise in antibiotic resistance, medical practitioners have turned to combination therapy with the hope of reducing the likelihood of new resistances to emerge. When two treatments act in such a fashion as to potentiate the efficacy of the other, the two are said to be synergistic. Harnessing such interactions results in a more robust treatment, and numerous studies have associated synergy with a reduction in the likelihood of antibiotic resistance emerging.

Just as two antibiotics can interact synergistically, the introduction of either genetic or epigenetic changes alongside standard drug treatments can elicit similar synergistic responses. Numerous CRISPR technologies have been developed to rapidly and efficiently affect such genetic and epigenetic changes. A thorough understanding of the potential synergy (and antagonism) between such modifications and available antibiotics would prove indispensable as it would serve as a powerful guiding force for the design of novel sequence-specific therapies to complement our current arsenal of antimicrobials.

Here, we establish a framework for interrogating such gene-drug synergy by employing gene knockouts and CRISPR interference of bacterial genes. We systematically explored the interactions between 30 specific gene knockouts and nine commonly used antibiotics, and show how this approach can be employed to rapidly characterize synergistic (and antagonistic) interactions. We extend these results to the gene expression landscape and demonstrate how employing deactivated CRISPR-Cas9 constructs to selectively inhibit gene expression replicated these synergistic interactions during simultaneous antibiotic exposure. We further show that combining gene expression inhibitions can exploit inherent epistatic interactions to amplify the degree of synergy with antibiotics. Finally, we show how this synergy potentiates antibiotic activity in an infection model, significantly reducing the bacterial load in HeLa cells infected with Salmonella enterica serovar Typhimurium.

These results demonstrate a new pipeline for finding and designing optimal sequence-specific drug targets that potentiate the activity of existing treatments. Synergy between gene-drug treatments can profoundly impact the ability of microorganisms to escape therapy, and these results highlight the efficacy of these strategies to combat the spread of antibiotic resistance

Development of CRISPR-Cas3 Weaponized Bacteriophages As Novel Antimicrobials

Conference: International Conference on CRISPR Technologies

Year: 2017

Proceeding: International Conference on CRISPR Technologies

Group: General Submissions

Session: CRISPR technologies beyond genome editing and gene regulation

Time: Tuesday, December 5, 2017 - 12:25pm-12:50pm

Authors:

Ousterout, D., Locus Biosciences

The rapid rise of antibiotic resistance has presented a significant challenge to combating bacterial disease and is projected to cause nearly 10 million deaths by 2050 if new approaches are not developed. This high unmet need has renewed interest in developing novel antimicrobials based on small prokaryotic viruses called bacteriophages. While phage therapy has long been associated with safety, therapeutic efficacy has yet to be demonstrated in well controlled clinical trials. Locus Biosciences is an emerging biotechnology company that is developing narrow spectrum antimicrobials to combat antibiotic resistant infections by exploiting the novel mechanism of action of CRISPR-Cas3 to trigger programmed cell death in targeted bacterial populations. CRISPR-Cas systems naturally act as adaptive immune systems in bacteria that can be readily programmed to target DNA for destruction. We have adapted Type I CRISPR-Cas3 systems as programmable exonucleases to destroy DNA efficiently, specifically, and beyond repair, thereby killing a targeted bacterium. Locus is pairing CRISPR-based killing technology with phages to create CRISPR-phages (crPhages) that act as DNA delivery vectors to achieve robust infection and sequence-specific programmed cell death in targeted bacterial populations. We will present work demonstrating proof-of-concept data for a crPhage product with activity against E. coliisolates in vitro and in vivo, as well as preliminary preclinical studies that support the use of designer crPhage products to eliminate bacterial diseases.

C-Berst: Defining Subnuclear Proteomic Landscapes at Genomic Elements with dCas9-APEX2

Conference: International Conference on CRISPR Technologies

Year: 2017

Proceeding: International Conference on CRISPR Technologies

Group: General Submissions

Session: CRISPR technologies beyond genome editing and gene regulation

Time: Tuesday, December 5, 2017 - 12:00pm-12:25pm

Authors:

Gao, X. D., University of Massachusetts Medical School
Tu, L. C., University of Massachusetts Medical School
Mir, A.
Dekker, J., University of Massachusetts Medical School
Shaffer, S. A., University of Massachusetts Medical School
Zhu, L. J., University of Massachusetts Medical School
Wolfe, S. A., University of Massachusetts Medical School
Sontheimer, E. J., University of Massachusetts Medical School

Spatial and temporal genome organization is essential to gene regulation and to all aspects of chromosome dynamics[1]. Thanks in large part to chromosome conformation capture (3C)[2], Hi-C, and related techniques, we are able to identify many physical interactions of entire genomes in four dimensions (4D). To further understand the specifications and interactions of the 4D nucleome, it is important to develop minimally biased, high-throughput methods for identifying the proteomic landscapes that shape and reflect subnuclear structure. Recently, nuclease-dead Cas9 (dCas9)-based techniques have enabled locus-specific genome labeling and gene regulation[3]. Separately, spatially restricted enzymatic tagging (SRET) approaches [such as those involving localized ascorbate peroxidase (APEX)-catalyzed biotinylation] have been applied successfully to map subcellular proteomes[4]. Here we report the development of dCas9-APEX2 Biotinylation at genomic Elements by Restricted Spatial Tagging (C-BERST) for the unbiased mapping of subnuclear proteomes in proximity to defined genomic loci in living U2OS cells[5]. In the presence of a single-guide RNA complementary to telomeric repeats (sgTelo), dCas9-APEX2 enables the specific biotinylation, enrichment and proteomic identification of 192 proteins via mass spectrometry. The striking success of our approach is exemplified by the six subunits of the telomeric shelterin complex, which were among the seven most highly enriched proteins that we identified. Another highly enriched protein was Apollo, a 5' to 3' exonuclease that interacts with the shelterin component TRF2 and functions in the alternative lengthening of telomeres (ALT) pathway that maintains telomere length in this cell type[6]. Overall, among the 192 most significantly sgTelo-enriched proteins, 32 have been reported previously to be associated with telomeres or linked to telomere function. These include proteins from complexes known to contribute to ALT-associated pathways or processes. Gene ontology analysis of the 192 C-BERST proteomic hits reveals strong functional associations with terms such as telomere maintenance, DNA replication, DNA repair, and homologous recombination, all of which are important for ALT pathways[7]. By combining the flexibility of RNA-guided dSpyCas9 genome binding with the efficiency and rapid kinetics of APEX2-catalyzed biotinylation, C-BERST extends the unbiased definition of subnuclear proteomes to many genomic elements, and to a range of dynamic processes that occur too rapidly to analyze via the 18-24h labeling procedures inherent to proximity-dependent biotin identification (BioID). Importantly, C-BERST promises to augment and extend Hi-C and related methods by linking conformationally important cis-elements with the factors that associate with them.

REFERENCES

1. Gorkin, D. U., Leung, D. & Ren, B. Cell Stem Cell 14, 762–775 (2014).

2. Dekker, J., Rippe, K., Dekker, M. & Kleckner, N. Science 295, 1306–1311 (2002).

3. Dominguez, A.A., Lim, W.A. & Qi, L.S. Nat Rev Mol Cell Biol 17, 5-15 (2016).

4. Rhee, H.-W. et al. Science 339, 1328–1331 (2013).

5. Gao, X. D. et al. Biorxiv 171819 (2017). doi: 10.1101/171819.

6. Lenain, C. et al. Curr Biol 16, 1303-1310 (2006).

7. Cesare, A.J. & Reddel, R.R. Nat Rev Genet 11, 319-330 (2010).

Using Cas9 for Gene Editing in Tribolium Castaneum: What Do We Need to Make It Work?

Conference: International Conference on CRISPR Technologies

Year: 2017

Proceeding: International Conference on CRISPR Technologies

Group: General Submissions

Session: Genome editing and gene regulation in agriculture

Time: Tuesday, December 5, 2017 - 1:40am-2:05am

Authors:

Grubbs, N.
Slayton, L.
Gutzmann, N.
Chu, F. C.
Lorenzen, M., North Carolina State University

The RNA-guided nuclease, Cas9, part of the bacterial adaptive immune system called CRISPR, has been hailed as a revolutionary tool for genome editing. As a result, it has been effectively deployed in several species to induce mutations in target genes, as well as insert genetic sequences of interest. The ability to identify and target any sequence of interest has made the CRISPR/Cas9 gene-editing system particularly attractive for research in species with limited resources, or for genomic stretches that have been resistant to other methods of study. However, outsideDrosophila melanogaster, little has been done to study the efficacy of the system in insects. Understanding how Cas9 function may vary from species to species, from gene to gene, and even from sequence to sequence will be important for this tool to be effectively applied in novel ways. Recently, researchers published the successful use of Cas9 for gene editing in Tribolium castaneum. In our efforts to use Cas9 for dissection of the selfish genetic element, Medea, in this model organism, we encountered several issues that were not described by these researchers, inspiring us to pursue a more detailed look at the efficiency and limitations of gene editing in Tribolium. I will discuss some of our discoveries, including an analysis of the impact of target sequence on guide efficacy, and how the relative positions of guide targets affects the use of multiple guides.

Production of High Oil, Transgene Free Camelina sativa Plants through Genome Editing

Conference: International Conference on CRISPR Technologies

Year: 2017

Proceeding: International Conference on CRISPR Technologies

Group: General Submissions

Session: Genome editing and gene regulation in agriculture

Time: Tuesday, December 5, 2017 - 1:15am-1:40am

Authors:

Jali, S., North Carolina State University
Hornstein, E., North Carolina State University
Edwards, B., North Carolina State University
Sederoff, H., North Carolina State University

Camelina sativa ('false flax,' Brassicaceae) is a traditional European oilseed that is now an engineering target as a dedicated bioenergy crop. Camelina oil's high unsaturated fatty acid content makes it uniquely suited for biodiesel production; however, low overall seed yield limits agricultural profitability. The hexaploid Camelina genome has been sequenced and efficient transformation is achieved through Agrobacterium-mediated gene transfer via floral dip. We have used the CRISPR/Cas9 gene editing system to target genes involved in the regulation of carbon assimilation, transport, and fatty acid biosynthesis pathways in order to increase seed yield and oil content. The use of CRISPR/Cas9 genome editing has allowed the generation of custom mutants for gene function studies, which are used in applied work on yield improvement for increased profitability of this crop. We have also established an amplicon deep-sequencing genotyping method to identify edited sites, which allows us to overcome both background SNPs in the allohexaploid plant and variation in CRISPR edits across plants. To date using this method as well as conventional methods we have genotyped up to the third transgenic generation of multiple lines, and are currently assessing yield and oil content.

Acknowledgement: This research is funded by BETO award from the Department of Energy in collaboration between North Carolina State University and Yield 10 (Formerly MetaboliX)

Livestock Improvements through Gene Editing

Conference: International Conference on CRISPR Technologies

Year: 2017

Proceeding: International Conference on CRISPR Technologies

Group: General Submissions

Session: Genome editing and gene regulation in agriculture

Time: Tuesday, December 5, 2017 - 12:00pm-12:25pm

Authors:

Cigan, M.

Improvements in livestock genetics through crossing and selection to produce high quality meat and increased milk production is a slow and resource intensive process. Animal agriculture stands to gain a huge leap forward with the adoption of new technologies like gene editing to improve disease resistance which will not only

improve the health and well-being of the animal but reduce the need for antibiotics. My talk will include gene editing projects at Genus that aim to improve animal health by delivering disease resistant pigs and cattle to farmers while maintaining food quality and enhancing sustainability.

Cms1 - a New Type V CRISPR Nuclease System

Conference: International Conference on CRISPR Technologies

Year: 2017

Proceeding: International Conference on CRISPR Technologies

Group: General Submissions

Session: Genome editing and gene regulation in agriculture

Time: Tuesday, December 5, 2017 - 12:00pm-12:25pm

Authors:

Begemann, M.
January, E.

The application of CRISPR genome editing technology is limited by the availability of nucleases capable of generating targeted double strand breaks across target organisms. The discovery of new nucleases with unique chemistry or divergent PAM sites expands the options for genome editing. We have identified a new Type V CRISPR system fromMicrogenomates and Smithella bateria (Cms1) that are capable of genome editing in eukaryotic systems. The Cms1 nucleases were originally shown to function in planta and their functionality in other eukaryotic systems is currently being explored. These nucleases are capable of cutting at an AT rich PAM site and do not require a tracrRNA. Multiple Cms1 nucleases are being characterized at a biochemical level to identify the most applicable nucleases across diverse target systems.

A CRISPR-Cpf1 System for Efficient Genome Editing and Transcriptional Repression in Plants

Conference: International Conference on CRISPR Technologies

Year: 2017

Proceeding: International Conference on CRISPR Technologies

Group: General Submissions

Session: Genome editing and gene regulation in agriculture

Time: Tuesday, December 5, 2017 - 12:25pm-12:50pm

Authors:

Qi, Y.

Clustered regularly interspaced short palindromic repeats (CRISPR)-Cpf1 has emerged as an effective genome editing tool in animals. Here we compare the activity of Cpf1 from Acidaminococcus sp. BV3L6 (As) andLachnospiraceae bacterium ND2006 (Lb) in plants, using a dual RNA Polymerase II promoter expression system. LbCpf1 generated biallelic mutations at nearly 100% efficiency at four independent sites in rice T0 transgenic plants. Moreover, we repurposed AsCpf1 and LbCpf1 for efficient transcriptional repression in Arabidopsis, and demonstrated reduction of miRNA159 transcription >10-fold. Our data suggest promising applications of CRISPR-Cpf1 for editing plant genomes and modulating the plant transcriptome.

RNA Targeting and Editing with CRISPR-Cas13

Conference: International Conference on CRISPR Technologies

Year: 2017

Proceeding: International Conference on CRISPR Technologies

Group: General Submissions

Session: Horizons of CRISPR biology

Time: Wednesday, December 6, 2017 - 12:50pm-1:15am

Authors:

Gootenberg, J., Broad Institute

RNA plays important and diverse roles in biology, but molecular tools to manipulate and measure RNA are limited. We demonstrate that RNA- targeting CRISPR effector Cas13 can be engineered for mammalian cell RNA knockdown, binding, and RNA editing. Cas13 can be heterologously expressed in mammalian and plant cells for targeted knockdown of either reporter or endogenous transcripts, targeted RNA binding for transcript imaging, and programmable RNA editing. Our results establish CRISPR-Cas13 as a flexible platform for RNA targeting with wide applicability for studying RNA in mammalian cells.

Rapid and Scalable Characterization of CRISPR Technologies Using an E. coli Cell-Free Transcription-Translation System

Conference: International Conference on CRISPR Technologies

Year: 2017

Proceeding: International Conference on CRISPR Technologies

Group: General Submissions

Session: Horizons of CRISPR biology

Time: Wednesday, December 6, 2017 - 12:50pm-1:15am

Authors:

Marshall, R., University of Minnesota
Maxwell, C., North Carolina State University
Collins, S., North Carolina State University
Luo, M. L., North Carolina State University
Jacobsen, T., North Carolina State University
Beisel, C. L., North Carolina State University
Noireaux, V., University of Minnesota

We present the use of E. coli cell-free transcription-translation systems (TXTL) to vastly improve the speed and scalability of CRISPR characterization and validation. Unlike prior approaches that require protein purification or live cells, TXTL can express active CRISPR machinery from added plasmids and linear DNA, and TXTL can output quantitative dynamics of gene repression. To demonstrate the applicability of TXTL, we rapidly measure guide RNA-dependent DNA cleavage and gene repression for single- and multi-effector CRISPR-Cas systems, accurately predict the strength of gene repression in E. coli, quantify the inhibitory activity of anti-CRISPR proteins, and develop a fast and scalable high-throughput screen for protospacer-adjacent motifs. These examples underscore the potential of TXTL to facilitate the characterization and application of CRISPR technologies across their many uses.

Phylogenomics of Cas4 Family Nucleases

Conference: International Conference on CRISPR Technologies

Year: 2017

Proceeding: International Conference on CRISPR Technologies

Group: General Submissions

Session: Horizons of CRISPR biology

Time: Wednesday, December 6, 2017 - 12:00pm-12:25pm

Authors:

Hudaiberdiev, S., Ncbi, Nlm, Nih

The Cas4 family endonuclease is a component of the adaptation module in many variants of CRISPR-Cas adaptive immunity systems. Unlike most of the other *cas* genes, *cas4* is often encoded outside CRISPR-*cas* loci (solo-Cas4) and is also found in mobile genetic elements (MGE-Cas4), particularly, archaeal viruses.

As part of our ongoing investigation of CRISPR-Cas evolution, we explored the phylogenomics of the Cas4 family. Notably, about 90% of the archaeal genomes encode Cas4 compared to only about 20% of the bacterial genomes. Many archaea encode both the CRISPR-associated form (Cas-Cas4) and solo-Cas4, whereas in bacteria, this combination is extremely rare. The solo-*cas4* genes are notably over-represented in environmental bacteria of the Candidate Phyla Radiation (CPR) that typically lack CRISPR-Cas systems, suggesting that Cas4 could be a component of uncharacterized defense systems of the CPR bacteria. The results of phylogenomic analysis indicate that both the CRISPR-associated *cas4* genes are often transferred horizontally but almost exclusively, as part of the adaption module, i.e. jointly with *cas1* and *cas2*genes. The evolutionary integrity of the adaptation module sharply contrasts the observed rampant shuffling of CRISPR-*cas* modules whereby a given variant of the adaptation module can combine with virtually any variety of effector modules. The solo-*cas4* genes evolve primarily via vertical inheritance and are subject only to occasional horizontal gene transfer. The estimated selection pressure on *cas4* genes does not substantially differ between the CRISPR-associated family members and solo-*cas4*, and is close to the genomic median. Thus, in terms of their evolutionary regime, *cas4* genes, similarly to *cas1* and*cas2*, behave like 'regular' microbial genes involved in various cellular functions, showing no evidence of direct involvement in virus-host arms races. A notable feature of the Cas4 family evolution is the frequent recruitment of *cas4* genes by various mobile genetic elements (MGE), particularly, archaeal viruses. The functions of Cas4 in these elements are unknown and potentially might involve anti-defense roles.

Unlike most of the other Cas proteins, Cas4 family members are as often encoded by stand-alone genes as they are incorporated in CRISPR-Cas systems. In addition, *cas4* genes were repeatedly recruited by MGE, perhaps, for anti-defense functions. Experimental characterization of the solo and MGE-encoded Cas4 nucleases is expected to reveal currently uncharacterized defense and anti-defense systems and their interactions with CRISPR-Cas systems.

Origins and Evolution of Tracrrnas in CRISPR-Cas Systems

Conference: International Conference on CRISPR Technologies

Year: 2017

Proceeding: International Conference on CRISPR Technologies

Group: General Submissions

Session: Horizons of CRISPR biology

Time: Wednesday, December 6, 2017 - 3:45am-4:10am

Authors:

Faure, G., NIH/NLM/NCBI
Shmakov, S.
Makarova, K. S.
Wolf, Y. I.
Koonin, E. V., NIH/NLM/NCBI

The CRISPR-Cas systems belong to two classes, with multi-subunit effector complexes in Class 1, and single multi-domain protein effectors in Class 2. The latter encompasses types II, V and VI that differ primarily by the domain architectures of the effector proteins. In all type II systems as well as subtype V-B, maturation of the precursor crRNA (pre-crRNA) requires a trans-acting CRISPR (tracr) RNA whereas in subtype V-A and type VI, no tracrRNA has been identified. In Cas9-CRISPR technologies, tracrRNA is artificially fused to crRNA to form a single-guide RNA. The tracrRNA consists of a portion complementary to the direct repeat from the corresponding CRISPR array and a unique portion. The origin and evolution of tracrRNA remain open problems which we sought to investigate by comparative analysis of co-folding between tracrRNA and crRNA. By comparing the available structures of Class 2 effectors complexed with the crRNA, tracrRNA (where involved) and the target, we identified the nexus structure of the tracrRNA in types II-A and II-B and a similar local fold in type V-B, where it involves both tracrRNA and the repeat portion of the crRNA, and in subtype V-A, where it is formed by the repeat alone. This observation suggests an important function of the nexus structure that is common to all type II and type V systems but could have evolved convergently, involving tracrRNA alone, the tracrRNA-repeat hybrid or the repeat alone. Using a larger set of tracrRNA-repeat pairs representative of the diversity of type II systems, we predicted the structures of each crRNA-tracrRNA co-folding and identified the nexus structure in almost all of them. The nexus is almost always located 2 to 3 nucleotides away from the tracrRNA-repeat hybrid region. We also observed that the 5'-terminal base of the repeat is always paired with the tracrRNA. We propose that the role of the nexus structure is to prevent base-pairing between tracrRNA and the spacer whereas the pairing of the 5' base of the repeat prevents interaction between the repeat and the DNA target. Thus, both features appear to ensure full availability of the spacer to interact specifically and completely with the DNA target. We investigated the origins of tracrRNA by tracing the coevolution of the tracrRNA and the corresponding repeats using simulated and ancestral RNA guide sequences, and observed a conservation of optimal hybrid energy that corresponds to partial, rather than complete, base-pairing. Then, we analyzed the tracrRNA in microbial genomes and found that tracrRNAs located close to CRISPR arrays are expressed from the same strand and downstream of the CRISPR array, and can even overlap with the array. We propose an evolutionary model in which tracrRNA emerges from the distal decaying repeat of the CRISPR array.

Nucleic Acid Detection with CRISPR-Cas13

Conference: International Conference on CRISPR Technologies

Year: 2017

Proceeding: International Conference on CRISPR Technologies

Group: General Submissions

Session: Horizons of CRISPR biology

Time: Wednesday, December 6, 2017 - 12:25pm-12:50pm

Authors:

Abudayyeh, O., Broad Institute/MIT

Rapid, inexpensive, and sensitive nucleic acid detection may aid point-of- care pathogen detection, genotyping, and disease monitoring. We combine the RNA-targeting CRISPR effector Cas13a with isothermal amplification to establish a CRISPR-based diagnostic (CRISPR-Dx), providing rapid DNA or RNA detection with attomolar sensitivity and single-base mismatch specificity. We use this Cas13a-based molecular detection platform, termed SHERLOCK, to detect specific strains of Zika and Dengue virus, distinguish pathogenic bacteria, genotype human DNA, and identify cell-free tumor DNA mutations.

Landscape of Governance for Gene Editing

Conference: International Conference on CRISPR Technologies

Year: 2017

Proceeding: International Conference on CRISPR Technologies

Group: General Submissions

Session: Horizons of CRISPR biology

Time: Wednesday, December 6, 2017 - 12:00pm-12:25pm

Authors:

Kuzma, J.

Gene editing is challenging existing regulatory systems and posing broader governance issues such as control of and access to the technology, public understanding and communication, and potential ease of use or misuse. In this presentation, I will review the current landscape of oversight for gene-edited products and recent policy proposals, as well as discuss emerging applications for CRISPR, such as in gene drives, and the special considerations they pose for governance.

Inhibition of Bacterial Immunity with Anti-CRISPR Proteins

Conference: International Conference on CRISPR Technologies

Year: 2017

Proceeding: International Conference on CRISPR Technologies

Group: General Submissions

Session: Horizons of CRISPR biology

Time: Wednesday, December 6, 2017 - 12:25pm-12:50pm

Authors:

Borges, A.
Zhang, J.
Watters, K. E., University of California, Berkeley
Rauch, B.
Doudna, J. A., Lawrence Berkeley National Lab
Bondy-Denomy, J., UCSF

Bacteria use CRISPR-Cas immune systems to defend against viral attack. In response, bacterial viruses have evolved a diverse suite of anti-CRISPR proteins that have previously been shown to inhibit (Type I) CRISPR-Cas3 systems in the bacterium Pseudomonas aeruginosa. To find anti-CRISPR proteins that inhibit (Type II) CRISPR-Cas9 systems, we developed a bioinformatics approach: Because anti-CRISPRs allow a CRIPSR targeted virus to integrate into the bacterial chromosome, we searched for genomes in which a Cas9 associated CRISPR array targeted a mobile element in the same genome, a circumstance we call "self-targeting". Using this approach, we discovered four unique CRISPR-Cas9-inhibitors encoded in Listeria monocytogenes integrated viruses. Given the success of this approach, we returned to examine CRISPR-Cas3 self-targeting in P. aeruginosa. We were surprised to see that greater than 10% (74 /713) of self-targeting events in P. aeruginosa could not be explained by the presence of known inhibitors, implying the existence of novel anti-CRISPRs. To find these novel inhibitors, we employ bioinformatics coupled with functional tests to uncover and characterize new strategies and mechanisms for CRISPR-Cas inhibition. These discoveries shed light on the interactions between CRISPR and bacterial viruses in the natural world.

A Bio-Nano Approach to Point-of-Care Testing of Parasitic DNA in Resource Limited Settings

Conference: International Conference on CRISPR Technologies

Year: 2017

Proceeding: International Conference on CRISPR Technologies

Group: Poster Session

Session: Poster Session

Time: Tuesday, December 5, 2017 - 4:05pm-6:35pm

Authors:

Bengtson, M., Technical University Delft
Pawlik, G.
Van de Torre, J.
Dekker, C., Kavli Institute of Nanoscience, Delft. Delft University of Technology

Effective treatment and prevention of infectious diseases requires accurate diagnosis. Current diagnostics are expensive, time consuming, require sophisticated equipment and adequate infrastructure – requirements that are often not met in resource limited settings. There is a great need for simple and cost effective point-of-care tests to probe for infectious diseases.

We aim to develop a point-of-care diagnostic test that probes for the presence of pathogenic DNA in body fluids. The key scientific innovation is to use the CRISPR/Cas9 genome-engineering system as a DNA detection tool. The Cas9 protein and its sensing guide RNA have the unique capability to find a specific target DNA sequence. Initial proof-of-principle experiments show that our CRISPR/Cas9 model is able to successfully detect the target DNA sequence. For readout, we have designed a Cas9-guide complex that will be coupled to an enzymatic DNA molecule (DNAzyme) that will facilitate a colorimetric readout in the presence of the target DNA sequence.

To increase sensitivity in detection, we are further developing an approach that simultaneously detects the target DNA sequence and amplifies the readout. A "trigger" DNA sequence, displaced in the presence of the target DNA, will initiate a cleavage-based amplification cycle that will result in the release of multiple DNAzyme molecules. The hybridization between the trigger and seed sequence in the absence of the target DNA ensures to avoid false-positive results.

Upon completion, the developed device is expected to be adaptable to detect any pathogenic DNA sequence, simply by equipping the guide RNA with a new target. The device will not only benefit those living in resource limited settings, but also provide a cost effective alternative to current diagnostics in developed countries.

[??1]What is the difference between these two events?

A High-Throughput Growth-Based Screen for CRISPR Pams Reveals Novel SpyCas9 Targeting Rules in Bacteria

Conference: International Conference on CRISPR Technologies

Year: 2017

Proceeding: International Conference on CRISPR Technologies

Group: Poster Session

Session: Poster Session

Time: Tuesday, December 5, 2017 - 4:05pm-6:35pm

Authors:

Slotkowski, R. A., North Carolina State University
Leenay, R. T., North Carolina State University
McGirr, B., North Carolina State University
Beisel, C. L., North Carolina State University

We developed and characterized a novel growth-based positive PAM assay to characterize CRISPR-Cas systems. This PAM screen manipulated the E. coli carbon metabolism so growth was only observed in the presence of D-xylose if a functional PAM was recognized on a replicated plasmid. The assay was characterized with the ubiquitousStreptococcus pyogenes Cas9 nuclease and instead of presenting its most functional PAMs, the screen enriched for weakly recognized NNGG PAMs. We demonstrated that S. pyogenes Cas9 is able to functionally target an NNGG motif using both transcriptional repression and plasmid clearance in a bacterial host. This PAM proved to be less functional than the canonical NGG motif, but still showed significant activity over non-PAMs. Altogether, we develped a new tool to discover the least function PAMs for CRISPR-Cas systems in bacteria and expanded the targeting rules for S. pyogenes Cas9.

Applying CRISPR for Genome Editing and Increasing Flux Towards Products in Cyanobacteria

Conference: International Conference on CRISPR Technologies

Year: 2017

Proceeding: International Conference on CRISPR Technologies

Group: Poster Session

Session: Poster Session

Time: Tuesday, December 5, 2017 - 4:05pm-6:35pm

Authors:

Gordon, G., Chemical and Biological Engineering, University of Wisconsin, Madison, WI
Courtney, D., Chemical and Biological Engineering, University of Wisconsin, Madison, WI
Korosh, T., Chemical and Biological Engineering, University of Wisconsin, Madison, WI
Cameron, J. C., University of Colorado-Boulder
Markley, A. L., University of Wisconsin-Madison
Pfleger, B., Chemical & Biological Engineering, University of Wisconsin, Madison, WI

CRISPR is a powerful tool for sequence specific selection and controlling gene expression, but implementation is not always straight forward even in bacteria. Using CRISPR paired with recombineering has become routine for gene knockouts, knock-ins, and precise genomic changes inEscherichia coli. We attempted to use CRISPR in industrially relevant fast growing cyanobacteria Synechococcus sp. strain PCC 7002, which could be used to turn carbon dioxide and sunlight into valuable products.We optimized a CRISPR interference system for tunable gene repression in cyanobacteria and were able to increase flux towards a desired product. By repression of a native nitrogen assimilation gene we were able to increase lactate production by 2-fold compared to a base engineered strain. Attempts for using CRISPR for genome editing in cyanobacteria has proved to be challenging. It has been reported that Cas9 is toxic to cyanobacteria, but we have made strains that express active Cas9. Initial results show that Cas9 seems to be cleaving the intended target, but we are unable to get the intended genomic edit. There seem to be specific challenges for using CRISPR for genome editing in some organisms including toxicity, reliance on specific DNA uptake mechanisms, lack of appropriate expression systems, and coordinated expression of both Cas9, guide RNA, and repair template.

Automated Bioinformatic Pipeline to Identify CRISPR-Cas Systems

Conference: International Conference on CRISPR Technologies

Year: 2017

Proceeding: International Conference on CRISPR Technologies

Group: Poster Session

Session: Poster Session

Time: Tuesday, December 5, 2017 - 4:05pm-6:35pm

Authors:

Crawley, A., NCSU
Barrangou, R., North Carolina State University
Henriksen, J.

While CRISPR-Cas systems have taken catapulted into the scientific spotlight due to their ability to be repurposed as genome editing tools, these systems natively act as adaptive immune systems in bacteria and archaea. Growth in the field has led CRISPR knowledge, publications, and nomenclature to progress at an overwhelming rate. We have developed an automated pipeline that is able to detect CRISPR repeats as well ascas genes and use these features to determine CRISPR-Cas Type and Subtype to aid CRISPR novices and experts in accelerated characterization of these systems. This pipeline accurately detects CRISPR loci and determines the completeness of systems present in bacterial and archaeal genomes. The pipeline uses cas genes presence and absence, as well as CRISPR repeat information, to accurately assign Types and Subtypes to candidate CRISPR-Cas systems. All six major types and 23 subtypes are detectable by this pipeline; in addition, the proteins in the putative V-U type are also detected. The pipeline can also be used to determine the novelty of proteins identified through homology searches to known databases. Additionally, users can use the pipeline to search for protein domains within sequences identified in the search. This tool will make reproducible, standardized, accessible, transparent and high-throughput analysis methods available to all researchers in and beyond the CRISPR-Cas research community. This is a useful and valuable tool as it enables classification within a complex nomenclature and provides analytical methods in field that is evolved rapidly.

Bacterial Genome Editing with CRISPR-Cas9: Deletion, Integration, Single Nucleotide Modification in Clostridium Beijerinckii as an Example

Conference: International Conference on CRISPR Technologies

Year: 2017

Proceeding: International Conference on CRISPR Technologies

Group: Poster Session

Session: Poster Session

Time: Tuesday, December 5, 2017 - 4:05pm-6:35pm

Authors:

Zhang, Z., UIUC

The application of the CRISPR-Cas9 system as a tool to genetically engineer bacteria represents an important improvement from previous methods, such as use of heterologous counter-selection markers, recombinases or the group II intron, all of which have limitations. Many industrially important bacteria are tough to work with due to the physiological characteristics they have adopted from their natural environment, such as thick cell walls, high GC content, spore formation and lack of sufficient chromosomal DNA repair mechanisms. This study demonstrates the use of a chimeric single RNA-guided *Streptococcus pyogenes* CRISPR-Cas9 system as a selection tool for desired mutants, from recombinants of a plasmid-borne editing template, in place of Cas9-directed repair, using *Clostridium beijerinckii* as an example. High efficiency of gene deletion, integration and single nucleotide modification was achieved. We further proposed and investigated an innovative two-step approach, which does not rely on direct availability of Protospacer Adjacent Motif sequences for genome editing. Based on these advances, additional studies using a similar strategy has been adapted and applied to other bacterial species and strains, supporting the potential for precise genome editing by this approach.

Casper: A Versatile CRISPR Associated Software for Pathway Engineering and Research

Conference: International Conference on CRISPR Technologies

Year: 2017

Proceeding: International Conference on CRISPR Technologies

Group: Poster Session

Session: Poster Session

Time: Tuesday, December 5, 2017 - 4:05pm-6:35pm

Authors:

Mendoza, B., University of Tennessee Knoxville
Trinh, C. T., University of Tennessee

The rapid expansion of research into the CRISPR (Clustered Regularly-Interspaced Short Palindromic Repeats) system requires a powerful software to guide precise genome editing beyond model organisms. Here we present CASPER (CRISPR Associated Software for PathwayEngineering and Research), a versatile, flexible integrated software platform, that provides end-users the ability to use CRISPR tools to manipulate non-traditional biological systems beyond model organisms with flexible use of a library of Cas endonucleases. Utilizing multiple algorithms for on- and off-target analysis of the guided RNA/Cas complex, CASPER achieves higher accuracy at predicting experimental results. Further, CASPER enables end-users to design guided RNAs to target multiple chromosome loci at once and across a consortium of organisms.

Controlling Transcription in the Yeast Yarrowia Lipolytica

Conference: International Conference on CRISPR Technologies

Year: 2017

Proceeding: International Conference on CRISPR Technologies

Group: Poster Session

Session: Poster Session

Time: Tuesday, December 5, 2017 - 4:05pm-6:35pm

Authors:

Schwartz, C., UC Riverside
Curtis, N.
Wheeldon, I., UC Riverside

The yeast Yarrowia lipolytica has been the subject of extensive metabolic engineering and synthetic biology studies due to its ability to make and store intracellular lipids. Recent engineering efforts have been aided by the development of a suite of different engineering tools, including well characterized promoters for heterologous gene expression, a Cre-lox recombinase system for selectable marker recycling, and CRISPR-Cas9-based tools for gene disruptions, integrations, and transcriptional repression. We have adapted our previously developed CRISPR-Cas9 system to enable the activation of transcription (CRISPRa). Different activation domains were systematically compared for their ability to drive expression of GFP from a model synthetic promoter, with the VPR domain giving the highest expression. A series of sgRNAs targeting the same model synthetic promoter were then designed, and co-expressed with a dCas9-VPR fusion protein. Fluorescence was measured, and CRISPRa activation of GFP was observed. The CRISPRa system was then tested on three different native Y. lipolytica promoters expressing GFP, and fluorescence was measured. This CRISPRa system will enable the activation of arbitrary native Y. lipolytica genes to facilitate the production of oleochemicals of interest.

CRISPR Antimicrobial Design for the Control of Bacterial Plant Pathogens

Conference: International Conference on CRISPR Technologies

Year: 2017

Proceeding: International Conference on CRISPR Technologies

Group: Poster Session

Session: Poster Session

Time: Tuesday, December 5, 2017 - 4:05pm-6:35pm

Authors:

Fagen, J., North Carolina State University
Lince, K.
Collias, D., North Carolina State University
Beisel, C. L., North Carolina State University
Braswell, E.

The prokaryotic adaptive immune system, CRISPR, has evolved to target mobile genetic elements, such as phage and plasmids, for destruction in a nucleotide sequence specific manner. Cell death is frequently observed when CRISPR systems are coerced into targeting the bacterial host's chromosome rather than invading DNA. Building on this capacity for induced autoimmunity, we have developed a CRISPR-based antimicrobial for highly specific bacterial killing. CRISPR antimicrobials are uniquely suited for removing individual strains or genes from complex communities and populations. This feature makes them highly desirable for treatment of bacterial pathogens in situations where broad-spectrum antimicrobials are impractical or undesirable. Plant agriculture in particular, is facing growing consumer rejection of traditional antimicrobial therapies concurrently with an increase in ecological pressures. Bacterial plant pathogens such as Liberibacter, Ralstonia, and Xanthomonas are threatening agricultural systems worldwide and growers need innovative crop protection solutions to manage these diseases. Here we investigate the development potential of CRISPR antimicrobials for mitigation of bacterial disease via direct plant application and treatment of pathogen reservoirs such as soils and irrigation water. We also demonstrate the capacity of CRISPR antimicrobials for target-specific killing with minimal perturbation of plant-associated microbial communities.

CRISPR Characterization and Diversity in Lactobacillus Fermentum

Conference: International Conference on CRISPR Technologies

Year: 2017

Proceeding: International Conference on CRISPR Technologies

Group: Poster Session

Session: Poster Session

Time: Tuesday, December 5, 2017 - 4:05pm-6:35pm

Authors:

Brandt, K., North Carolina State Unviersity
Barrangou, R., North Carolina State University

Lactic Acid Bacteria are gaining notoriety not only for their historical food benefits, but most recently for their rising potential in the microbiome, as probiotics, and development of CRISPR technology. Lactic Acid Bacteria are enriched for CRISPR systems, especially the Type II-A CRISPR-Cas9 system. With a growing desire to utilize CRISPR technologies for genome editing, it has become evident that the CRISPR toolbox will have to expand. Part of the Lactic Acid Bacteria complex, Lactobacillus fermentum is typically found in food products and human metagenome studies. It has recently increased in popularity due to its potential probiotic and antimicrobial effects. Due to these reasons, we investigated and characterized the CRISPR-Cas systems of Lactobacillus fermentum. Mining for CRISPR systems in 30 publicly available genomes, revealed a rich diversity of systems within this species. We determined the occurrence of CRISPR across genomes, and assigned Class and Type for each candidate. 70% of the genomes contain a putative CRISPR-Cas system, of which 81% contain a Type I CRISPR-Cas system and 52% encode a Type II CRISPR-Cas system. Our findings showed that while most systems did have a CRISPR locus, the number and type could differ greatly across genomes within this species. Next, we looked at the similarity of Cas1 and Cas2 sequences between strains. Predictably, the groupings were consistent based on the type of system, rather than the strain. Next, we examined the spacer profile. Contrary to previous studies, we found that the spacer profile of each system varied more than expected, with no conserved ancestral spacers. Actually, only two groups shared common ancestors, while the others showed unique spacer acquisition tracks. Finally, we investigated the Cas9 from the model strain ATCC 14931, a distant homolog of Sth and Spy canonical Cas9 proteins sharing only 32% AA identify, as a potential orthogonal system for the development of novel CRISPR-based molecular tools.

CRISPR-Induced White Eyes in Tribolium Castaneum

Conference: International Conference on CRISPR Technologies

Year: 2017

Proceeding: International Conference on CRISPR Technologies

Group: Poster Session

Session: Poster Session

Time: Tuesday, December 5, 2017 - 4:05pm-6:35pm

Authors:

Oppert, B., USDA ARS CGAHR
Adrianos, S., USDA ARS CGAHR
Lorenzen, M., North Carolina State University

White-eyed insects are nothing new; insects with normally dark eyes are known to have spontaneous mutants with white eyes. In insects with eyes solely pigmented by ommochromes, the metabolic pathway begins with a gene called *vermilion* that encodes the enzyme tryptophan 2,3-dioxygenase, and white-eyed insects also have been obtained through disruption of this gene. We explored the use of CRISPR to disrupt the*vermilion* gene in the red flour beetle, *Tribolium castaneum*, as a first step in transferring the technology to other stored product insects to produce white-eyed mutants. Guide RNA was targeted to *T. castaneum*vermilion, and newly-laid *T. castaneum* eggs were injected with a mixture of gRNA and Cas9 protein. Adult survivors had black eyes and were mated to a mutant strain of *T. castaneum* with white eyes, Tcv^{w}, in which the mutant *vermilion* gene contains a deletion of promoter and 80% of coding sequence. One CRISPR Go survivor mated to a Tcv^{w}individual had 19% white-eyed G1 progeny, and sequencing confirmed that there were multiple CRISPR events. Some of the G1 progeny with black eyes had a six base pair deletion around the gRNA site, which resulted in an in-frame deletion and a loss of two amino acids in the tryptophan 2,3-dioxygenase enzyme. However, sequencing also revealed that white-eyed G1 progeny had two base pair deletions causing a frameshift and insertion of a premature stop codon, with presumably no enzyme product. The distinction of the CRISPR allele (Tcv^{CRw}) and mutant allele (Tcv^{w}) in white-eyed G1 progeny was further confirmed via PCR using diagnostic primers specific for each allele. These data confirm the utility of the CRISPR system to produce eye color mutants in beetles for efficient markers in downstream germline transformation projects.

Determining on- and Off-Target Excision Events of Anti-HIV-1 Grnas Used Infor CRISPR/Cas9-Mediated Excision Therapy

Conference: International Conference on CRISPR Technologies

Year: 2017

Proceeding: International Conference on CRISPR Technologies

Group: Poster Session

Session: Poster Session

Time: Tuesday, December 5, 2017 - 4:05pm-6:35pm

Authors:

Sullivan, N., Center for Molecular Virology and Translational Neuroscience
Dampier, W., Drexel University College of Medicine
Homan, G., Center for Molecular Virology and Translational Neuroscience, Institute for Molecular Medicine and Infectious Disease,
Allen, A., Drexel University College of Medicine, Philadelphia
Chung, C. H., Drexel University College of Medicine
Pirrone, V., Center for Molecular Virology and Translational Neuroscience
Passic, S., Center for Molecular Virology and Translational Neuroscience
Williams, J., Center for Molecular Virology and Translational Neuroscience
Zhong, W., Drexel University College of Medicine
Kercher, K., Center for Molecular Virology and Translational Neuroscience
Szep, Z., Drexel University College of Medicine
Jacobson, J., Temple University
Nonnemacher, M., Drexel University College of Medicine
Wigdahl, B., Center for Molecular Virology & Translational Neuroscience

While antiretroviral therapy (ART) has reduced viral loads to undetectable levels, in most patients, it does not remove the integrated HIV proviral DNA from the latent reservoirs, which is athe major barrier to a cure. The CRISPR/Cas9 gene-editing system has been shown to excise the entire HIV-1 genome by targeting the long terminal repeats (LTRs). However, published anti-HIV-1 gRNAs have not been shown to effectively target the LTRs of the viral quasispecies within or between infected individuals. Our in silico analysis has demonstrated only two previously published anti-HIV-1 gRNAs targeting the LTR can cut >90% of the patient-derived

HIV-1 subtype B LTRs from the *Los Alamos National Laboratory* (LANL) HIV database. To improve on these results next-generation sequencing was performed on 269 patients enrolled in the Drexel Medicine CNS AIDS Research and Eradication Study (CARES) Cohort to understand and design gRNAs to account for thevariation in the HIV-1 LTR . A novel gRNA design pipeline was designed to select for optimal gRNAs and filter out those with predicted off-target binding. These results informed the design of a 4 and 10 selected molecular gRNA targets (SMRT) package. Utilizing an in silico experimental approach, the SMRT-4 and -10 packages are able to excise all subtype B LTRs from LANL and the CARES Cohort, without any predicted off-target excision. Functional studies have demonstrated that the SMRT 4 gRNAs performedperforming better at reducing expression and mutating the LTR than other published gRNAs in a number of different assays. However, few studies have fully examined the potential off-target excision by anti-HIV-1 gRNAs. Therefore, to determine if there is off-target excision, genome-wide, unbiased identification of DSBs enabled by sequencing (GUIDE-Seq) will be used to determine on- and off-target excision of SMRT gRNAs and others on a whole-genome level with NGS sequence coverage. Overall, these studies will advance a second-generation of CRISPR/Cas9 therapeutics and inform the design and testingdesing of novel anti-HIV-1 gRNAs.

Developing CRISPR-Cas-Based Genome Editing Tools for High-Throughput Engineering of Non-Model Prokaryotes

Conference: International Conference on CRISPR Technologies

Year: 2017

Proceeding: International Conference on CRISPR Technologies

Group: Poster Session

Session: Poster Session

Time: Tuesday, December 5, 2017 - 4:05pm-6:35pm

Authors:

Bosma, E. F., DTU Biosustain
Aragão Börner, R., Technical University of Denmark
Lennen, R., Technical University of Denmark
Nielsen, A. T., Technical University of Denmark

Although model organisms such as *Escherichia coli* and *Saccharomyces cerevisiae* have shown to be efficient industrial work horses for production of green chemicals and fuels from biomass, the use of alternative production hosts could provide important operational advantages to create more sustainable and economically feasible processes[1]. However, the development of such non-model organisms into cell factories is hampered by a lack of knowledge on their metabolism and underdeveloped or absent genetic tools for the required metabolic engineering[1,2]. Established methods for genomic modifications such as recombineering or plasmid-based homologous recombination (HR) have a high wild-type revertant rate, resulting in low engineering efficiencies, thus requiring extensive screening or insertion of markers. This also limits the use of these methods in organisms with low transformation and recombination efficiencies. Contrary to eukaryotes, most prokaryotes do not have a functional non-homologous end-joining system and double-strand DNA breaks (DSBs) induced by Cas9 (or other Cas-like molecules) are lethal, which forms the basis for using CRISPR-Cas-systems in prokaryotes as powerful counter-selection system when combined with a recombination method[3]. This has strongly increased engineering efficiencies in *E. coli* as well as several non-model organisms[2-4]. A major issue for non-model organisms is the lack of genetic parts such as plasmids, markers and tightly controllable promoters. Therefore, the Cas9-based system should be as simple as possible. Recently, a*Streptococcus pyogenes* (sp)Cas9-based method was developed in the facultative thermophile *Bacillus smithii*, based on a single plasmid containing HR-flanks, the gene coding for spCas9 and an sgRNA targeting the wild-type sequence selected for mutagenesis via HR[4]. The system does not need inducible promoters as it was found that spCas9 is inactive*in vivo* ≥42°C, and its activity can thus be controlled via the growth temperature[4]. This enabled highly efficient genome engineering by transformation and HR at elevated temperatures when spCas9 is inactive, and subsequent counter-selection of non-edited cells at 37°C by active spCas9[4]. Here, we apply this method in other non-model organisms to extend the range of bacteria for which this efficient Cas9-based counter-selection can be used. This also includes optimization or establishment of HR-methods. Altogether, by further developing the genome editing toolbox for these organisms we aim to broaden the range of potential cell factories to species that are for example more tolerant to low pH, high temperature and toxic substrates and products.

References

1. Bosma, E. F., van der Oost, J., de Vos, W. M. & van Kranenburg, R. Sustainable production of bio-based chemicals by extremophiles. Curr. Biotechnol. 2, 360–379 (2013).
2. Ronda, C., Pedersen, L. E., Sommer, M. O. A. & Nielsen, A. T. CRMAGE: CRISPR Optimized MAGE Recombineering. Sci. Rep. 6, 19452 (2016).
3. Mougiakos, I.*, Bosma, E.F.*, de Vos, W.M., van Kranenburg, R., van der Oost, J., Next generation prokaryotic engineering: The CRISPR-Cas toolkit. Trends Biotechnol. 34, 575-87 (2016).
4. Mougiakos, I.*, Bosma, E. F.*, Weenink, K., Vossen, E., Goijvaerts, K., van der Oost, J., and van Kranenburg, R. Efficient genome editing of a facultative thermophile using the mesophilic spCas9. ACS Synth. Biol. 6, 849-861 (2017).

Development of Tools for Precise Genome Engineering in Lactococcus Lactis

Conference: International Conference on CRISPR Technologies

Year: 2017

Proceeding: International Conference on CRISPR Technologies

Group: Poster Session

Session: Poster Session

Time: Tuesday, December 5, 2017 - 4:05pm-6:35pm

Authors:

Dudnik, A., Technical University of Denmark
Förster, J., Technical University of Denmark
Norholm, M. H. H., The Novo Nordisk Foundation Center for Biosustainability, Technical University of Denmark

Strains of Lactic Acid Bacteria (LAB) have a broad range of applications in food industry, including manufacturing of cheese, sausages, and other fermented products. Due to their robustness and stress tolerance, LAB are also being explored as potential candidates for production of fine chemicals within a frame of several projects, including BacHBerry. The latter is focused on identification and production of novel high-value plant-borne polyphenolic compounds, such as flavonoids using bacterial cell factories. Many LAB strains are easily genetically accessible, as there exist efficient transformation protocols, as well as expression vectors, and classical tools for genome modification. Major drawbacks of currently available genome modification strategies are that they are time consuming, require several rounds of selection, and in most cases only a single locus can be targeted at a time. Recent studies were able to overcome these issues by using CRISPR/Cas (Clustered, Regularly Interspaced, Short Palindromic Repeats – CRISPR-associated proteins) and recombineering (recombination-mediated genetic engineering). The main goal of this project is to assemble a toolbox for rapid engineering ofLactococcus lactis, a well-studied species of LAB, which could be used for construction of efficient production strains. We plan to introduce and optimize a system for genome engineering of L. lactis in order to have a tool for rapid site-directed mutagenesis, as well as insertion and deletion of genetic elements. The most up-to-date results of the study would be discussed herein.

Disrupting Bacterial Adaptive Resistance Using dCas9 and dCas9-? Gene Perturbations: Emerging CRISPR Applications in Synthetic Biology

Conference: International Conference on CRISPR Technologies

Year: 2017

Proceeding: International Conference on CRISPR Technologies

Group: Poster Session

Session: Poster Session

Time: Tuesday, December 5, 2017 - 4:05pm-6:35pm

Authors:

Otoupal, P., University of Colorado Boulder
Erickson, K., University of Colorado Boulder
Bordoy, A. E., University of Colorado Boulder
Chatterjee, A., University of Colorado Boulder

The rate at which new antimicrobial resistances evolve has surpassed our ability to develop novel therapeutics to combat them. This has engendered a crisis that necessitates a greater understanding of how resistance emerges. Recent studies have suggested that, when exposed to antimicrobials, bacteria enter an "adaptive resistance" state by exploring multiple pathways sampling a dynamic gene regulatory space.

We investigated the impact of non-genetic factors in bacterial adaptive pathways by controllably up-regulating and down-regulating the expression of genes known to be involved in tolerance. To this end, we employed emerging synthetic biology techniques to investigate gene regulatory networks involved in controlling adaptive resistance. Using deactivated CRISPR-Cas9 (dCas9) and dCas9 fused with the omega subunit of RNA polymerase (dCas9-ω), we designed and characterized a library of synthetic genetic devices to activate and inhibit native gene expression of stress-response networks.

Here we demonstrate the effect these constructs have on Escherichia coliphenotypes during exposure to sub-minimal inhibitory concentrations of a range of toxins, including antibiotics and disinfectants. We show that significant control over bacterial fitness and growth phenotypes can be achieved, reducing (or improving) the cells' ability to proliferate during stress. The range of adaptive phenotypes observed allowed us to identify individual genes important in the evolution of bacterial tolerance towards specific stress conditions. We ultimately discern genes responsible for the development of bacterial tolerance, providing direction for novel therapeutics which can impede intrinsic adaptive pathways leading to antimicrobial resistance.

Efficiency and Outcomes of Sequence-Specific Self-Targeting in Streptococcus Thermophilus Utilizing Native CRISPR-Cas Type I-E and Type II-a Systems

Conference: International Conference on CRISPR Technologies

Year: 2017

Proceeding: International Conference on CRISPR Technologies

Group: Poster Session

Session: Poster Session

Time: Tuesday, December 5, 2017 - 4:05pm-6:35pm

Authors:

Canez, C., North Carolina State University
Barrangou, R., North Carolina State University

CRISPR-Cas systems have gained recognition due the specificity of their RNA-mediated, nuclease-dependent targeting, to provide adaptive immunity against phages. This machinery has been repurposed in eukaryotes which has revolutionized the means and potential for genome editing. These systems hinge on Cas effector nucleases to target and generate sequence-specific DNA damage via a double-stranded DNA break (Cas9) or substantial degradation of ssDNA (Cascade-Cas3). Due to notoriously limited DNA repair pathways in bacteria, CRISPR-Cas systems can be repurposed for lethal sequence-specific targeting, as an innovative form of precision antimicrobials. Our objective was to compare the relative killing efficiencies of endogenous Type I and Type II CRISPR-Cas systems in the model organism Streptococcus thermophilusDGCC7710. The hypothesis was that endogenous targeting via Cas3-mediated exonucleolytic DNA damage would elicit more deleterious effects than Cas9-directed endonucleolytic processing. Targeting of lacZby delivering plasmids carrying CRISPR arrays with targeting spacers repurposing either the native Type I-E or Type II-A systems. We observed efficient killing in both cases, in a dose-dependent manner when delivering 0.4-400ng of plasmid DNA. We next characterized survivors from both targets via PCR screening and sequencing to determine the genetic basis enabling survival. Results showed that Type I-E survival was primarily the result of low-frequency targeting-defective plasmids that lacked the targeting spacer, likely generated via loss through homologous recombination between CRISPR repeats. In contrast, Type II-A survivors had mutations in the chromosomal target. Surprisingly, we observed the excision of a 35 kB genomic segment, putatively generated via homologous recombination between flanking UDP-glucose-4-epimerase sequences (~250 bp identity) to circumvent self-targeting. Future efforts will determine whether Type I-based targeting generates more extensive damage that could yield more potent antimicrobials.

Engineering Cupriavidus Sp. for Ethylene Production Using CRISPR/Cas Systems

Conference: International Conference on CRISPR Technologies

Year: 2017

Proceeding: International Conference on CRISPR Technologies

Group: Poster Session

Session: Poster Session

Time: Tuesday, December 5, 2017 - 4:05pm-6:35pm

Authors:

Vaud, S., University of Nottingham
Bryan, S., University of Nottingham
Heeb, S., University of Nottingham
Winzer, K., University of Nottingham
Twycross, J., University of Nottingham

The chemoautotrophic bacterium Cupriavidus necator H16 (formerly known as *Ralstonia eutropha* H16) can utilise H_2 and CO_2 as a sole energy and carbon source to grow under aerobic conditions. It also accumulates large amounts of polyhydroxybutyrate, a polyalkanoate, used to manufacture biodegradable plastics. This microorganism is of great industrial interest, as metabolic engineering could lead to the production of other highly valuable compounds directly from industrial gas waste. One of these compounds is ethylene, global demand for ethylene is expanding at a rapid rate with production predicted to reach 200 million tonnes by 2020. Ethylene is currently produced from stream cracking or dehydrogenation of ethane, both of these produce large amounts of CO_2, which has a deleterious impact on the environment.We must find environmentally friendly alternatives to satisfy our energy requirements. We aim to engineer Cupriavidus sp. as a platform for the production of hydrocarbon-based products such as ethylene. High throughput metabolic engineering requires fast, flexible and user-friendly genome editing, these tools are not currently readily available inCupriavidus sp. We demonstrate that ethylene can be produced inCupriavidus metallidurans utilising the efe gene (ethylene forming enzyme) from P. syringae pv. paseolicola, which is sufficient for ethylene production in heterologous hosts. To improve productivity we are developing HTP genomic engineering tools utilising CRISPR/Cas9 and homologous recombineering.

Harnessing Cpf1 for CRISPR-Based Antimicrobials

Conference: International Conference on CRISPR Technologies

Year: 2017

Proceeding: International Conference on CRISPR Technologies

Group: Poster Session

Session: Poster Session

Time: Tuesday, December 5, 2017 - 4:05pm-6:35pm

Authors:

Singh, A.
Beisel, C. L., North Carolina State University

Bacteria and archaea have a unique adaptive immune system called CRISPR-Cas whose RNA-guided nucleases have been used for applications ranging from genome editing and gene regulation to in vitro diagnostics and DNA imaging. The Type II CRISPR effector protein Cas9 has been the most thoroughly investigated RNA-guided DNA endonuclease and was the pioneer for genome editing and gene regulation applications. More recently, Cas9 has been applied as an antimicrobial, where it has been successfully used to target antibiotic-resistance genes encoded on plasmids or in the genome as well as non-essential chromosomal genes. Unfortunately, Cas9 exhibits a low level of off-target toxicity and cleavage by Cas9 is not always lethal which limits further development as a targeted antimicrobial. Recently, new types of CRISPR nucleases have been discovered that may better serve as discretionary antimicrobial agents. The Type V-A CRISPR effector protein Cas12a (Cpf1) functions similarly to Cas9 but has several advantages over Cas9. Cpf1 is smaller in size and requires only single RNA for targeting the bacterial genome. Cpf1 has been shown to be highly specific with lower off target killing than Cas9 antimicrobials. Here, we examine the potency of Cpf1 as a CRISPR-based antimicrobial. This antimicrobial functioned even in the presence of the RecA DNA repair pathway. Subsequent cleavage led to cell death and is not dependent on essential gene targets. We further demonstrate the utility of this antimicrobial against E. coliO9:HS, Salmonella enterica and Klebsiella pneumoniae. These results indicate that Cpf1 may be a highly desirable effector for future development of CRISPR antimicrobials.

How Genome Editing Will Improve Agriculture

Conference: International Conference on CRISPR Technologies

Year: 2017

Proceeding: International Conference on CRISPR Technologies

Group: Poster Session

Session: Poster Session

Time: Tuesday, December 5, 2017 - 4:05pm-6:35pm

Authors:

Sirere, G. II, Technical University of Mombasa

HOW GENOME EDITING WILL IMPROVE AGRICULTURE.
New genetic techniques are improving our understanding of biological processes and genetic conditions, and are proving to be the best treatment options and disease prevention methods. With the aid of genome editing, areas of the genome linked to specific traits can now be precisely edited. Genome editing can provide a catalog of options for farmers to order exactly what they need. In crops, the technology has the potential to improve drought tolerance, eliminate diseases, increase yields, and much more. There are endless possibilities. With gene editing, the ability to pick livestock traits will be easy.

Gene editing allows scientists to engineer organisms without inserting transgenic DNA. Therefore, genome editing has a vast difference with Genetically Modified Organisms **(GMOs).**

Using these techniques to improve crops and livestock will contribute to address the pressing global challenge of food security.

ANIMAL WELFARE

People may be open to genetically engineered animals if it means more humane treatment, such as dairy calves that no longer require painful dehorning. Randall Prather, distinguished professor of animal sciences at the University of Missouri and director of the National Swine Resource and Research Center, helped develop pigs resistant to the deadly PRRS virus using CRISPR technology.

You change a single gene that allows the cow to thermoregulate better in heat. It is precision breeding. It is not science fiction.

â€œThis could have a significant impact on animal welfare,â€ says Prather. â€œNobody likes to see animals suffer.

â€œThere are physiological and emotional costs of these diseases, as well as economic, when they hit family farms,â€ he explains. â €œWhen I give talks about PRRS, I look out in the audience and see a wife pulling close to her husband, leaning in and tearing up. When I see that, I know those people know exactly what Iâ€™m talking about because it happened to them.â€

Scientists at the University of Edinburghâ€™s Roslin Institute are taking genes from warthogs resistant to African swine fever and inserting them into domesticated swine in an attempt to eventually eliminate this catastrophic disease from the earth. â€œThatâ€™s food security,â€ says Prather.

Implementation and Gene Editing of the Microorganism Rhodopseudomonas Palustris Trough the CRISPR Cas9 System

Conference: International Conference on CRISPR Technologies

Year: 2017

Proceeding: International Conference on CRISPR Technologies

Group: Poster Session

Session: Poster Session

Time: Tuesday, December 5, 2017 - 4:05pm-6:35pm

Authors:

Tapia, M. A., IPN
García-Peña, E. I., Instituto Politécnico Nacional
Salgado Manjarrez, E., Unidad Profesional Interdisciplinaria de Biotecnología, Instituto Politecnico Nacional
Jurado Marbán, V. H., Unidad Profesional Interdisciplinaria de Biotecnología, Instituto Politecnico Nacional

This work shows the application and implementation of the CRISPR Cas9 method to edit the genome of a microorganism that lacks the validated PAM and tracrRNA sequences required for the CRIPSR Cas9 gene editing tool (Ma et al, 2015). The purple non-sulfur bacteria used in this work is the highly adaptable Rhodopseudomonas Palustris, a microorganism capable of shifting between the four modes of metabolism that support life. This on itself, gives the microorganism a remarkable potential given the large amount of high value metabolites it can generate. It can produce hydrogen (H2), polyhydroxyalkanoates (PHA) and pigments, which are promising given their vast amount of applications and value (Laimer et al,2004). However as mentioned by other authors, the extraordinary adaptability the microorganism presents can become a major drawback, as it becomes exceedingly difficult to redirect the metabolism in a highly specific way (Kim et al, 2012; Natha et al, 2009; Wu et al, 2012; Lee et al, 2002). for instance, R. Palustris relies on the reducing power available to produce several metabolites and the metabolic routes for these often compete among them, the distribution of the reducing power promotes a competition between the production of PHB and hydrogen, or growth (Kim et al, 2012; McKinlay et al, 2013). Considering the previously stated, genetic manipulation of the microorganism is interesting since it is possible to maximize the production of a specific metabolite while reducing or eliminating the undesired. In order to perform an efficient and intelligent modification of the microorganism, the main metabolic routes involved in the production of the metabolites must be closely analyzed, in accordance to this point several authors report the performance of mutant strains with one or several modifications to said metabolic pathways(H_2, PHB and pigments) (Chen et al, 2012; Heiniger et al, 2011; Laguna et al, 2011) this information will facilitate the selection of the genetic modifications; also by realizing a gene expression analysis on R. Palustris during different culture conditions is possible to increase the efficacy of the overall process. To edit the genome of the microorganism the type II CRISPR cas9 system was selected due to the minimalist architecture of the mechanism and the elevated likelihood of Rhodopseudomonas having the type II-C CRISPR cas9 system (Burstein et al 2017).

Institutionalization of CRISPR/Cas9 through Collaboration: A Sociological Perspective on Gene Editing

Conference: International Conference on CRISPR Technologies

Year: 2017

Proceeding: International Conference on CRISPR Technologies

Group: Poster Session

Session: Poster Session

Time: Tuesday, December 5, 2017 - 4:05pm-6:35pm

Authors:

Molina, S.

This paper analyzes the spread of CRISPR technologies through a sociological lens. What are the practices and social patterns underlying researchers' capacity to multiply the applications of CRISPR technology? Moreover, how do these new applications become transferable and reliably reproduced in different laboratories; that is, how do they become institutionalized?

In order to address these questions, bibliometric analysis and observational research were conducted. To first identify general patterns and attributes of the social networks underlying the advancement of CRISPR technology, citation and co-authorship patterns from published articles and reviews are analyzed. This analysis identifies key papers and authors that were central to the spread of CRISPR/Cas9 and offers metrics about density and modularity of the network. On their own these networks does not explain the social mechanisms by researchers develop and adopt new applications of CRISPR/Cas9. To gain purchase on this, longitudinal and descriptive data was collected over the course of two years from observational research at laboratories in the San Francisco Bay Area. This ethnographic research involved attending regular lab meetings and conferences, listening in on meetings with collaborators, shadowing researchers on the lab bench, and interviews with researchers and staff in academic organizations.

The data from these observations suggests that in the process of entering new collaborative relationships between research programs, researchers have the opportunity to exchange tacit knowledge about how to successfully put new applications of CRISPR/Cas9 into practice. These opportunities are particularly salient when collaboration occurs between scientific research programs from different fields and problem areas. The results from this study suggests that without an understanding of how scientists develop standards of practice through collaboration, ongoing efforts by scientists, ethicists and policy makers to regulate the applications of CRISPR/Cas9 through bureaucratic means are incomplete.

Managing the SOS Response for Enhanced CRISPR-Cas-Based Recombineering in E. coli through Transient Inhibition of Host Reca Activity

Conference: International Conference on CRISPR Technologies

Year: 2017

Proceeding: International Conference on CRISPR Technologies

Group: Poster Session

Session: Poster Session

Time: Tuesday, December 5, 2017 - 4:05pm-6:35pm

Authors:

Moreb, E., Duke University
Lynch, M., Duke University
Menacho-Melgar, R., Duke University
Hoover, B., Duke University
Yaseen, A., Duke University
Roecker, Z., Duke University
Valyasevi, N., Duke University

Phage-derived "recombineering" methods are utilized for bacterial genome editing. Recombineering results in a heterogeneous population of modified and unmodified chromosomes, and therefore selection methods, such as CRISPR-Cas9, are required to select for edited clones. Cells can evade CRISPR-Cas-induced cell death through *recA*-mediated induction of the SOS response. The SOS response increases RecA dependent repair as well as mutation rates through induction of the *umuDC* error prone polymerase. As a result, CRISPR-Cas selection is more efficient in *recA*mutants. We report an approach to inhibiting the SOS response and RecA activity through the expression of a mutant dominant negative form of RecA, which incorporates into wild type RecA filaments and inhibits activity. Using a plasmid-based system in which Cas9 and recA mutants are coexpressed, we can achieve increased efficiency and consistency of CRISPR-Cas9-mediated selection and recombineering in *E. coli*, while reducing the induction of the SOS response. To date, this approach has been shown to be independent of *recA* genotype and host strain lineage. Using this system, we demonstrate increased CRISPR-

Cas selection efficacy with over 10,000 guides covering the *E. coli* chromosome. The use of dominant negative RecA or homologues may be of broad use in bacterial CRISPR-Cas-based genome editing where the SOS pathways are present.

Metabolic Engineering of Clostridium sp. to Increase Hydrogen Production Using the CRISPR/Cas System

Conference: International Conference on CRISPR Technologies

Year: 2017

Proceeding: International Conference on CRISPR Technologies

Group: Poster Session

Session: Poster Session

Time: Tuesday, December 5, 2017 - 4:05pm-6:35pm

Authors:

Soto Santibáñez, H., Unidad Profesional Interdisciplinaria de Biotecnología, Instituto Politécnico Nacional
Salgado Manjarrez, E., Unidad Profesional Interdisciplinaria de Biotecnología, Instituto Politecnico Nacional
García Pena, I., Unidad Profesional Interdisciplinaria de Biotecnología, Instituto Politecnico Nacional

Biofuels are eco-friendly since they do not contaminate the environment nor contribute to global warming, but present the disadvantage that their production costs are higher than those for fossil fuels and sometimes compete for food resources.

To reduce fuel-associated production costs, organic wastes can be used to produce biogas by microbial fermentation. A promising biogas is hydrogen since it is the fuel with the highest energetic capacity known and its combustion product is just water. Anaerobic Clostridia are potential producers of this fuel, and biotechnological techniques can be applied to improve fuel production as is the case of metabolic engineering. Previously, Gonzalez-Garcia et al. (2017) proposed a metabolic model of a mixed culture, containing Clostridium sp., which suggests that a theoretical yield as high as 10 mol H_2/mol of substrate can be achieved if H_2 pressure is sufficiently low. The same study suggests that H_2 yields can be improved by redirecting the metabolic flux towards the pentose phosphate pathway instead of glycolysis, thus increasing the reduction power necessary for a higher hydrogen production. Knocking out competing reactions for byproduct production are also suggested.

CRISPR/Cas system is a novel genome editing approach that has been widely used in several organisms and research areas. It offers various advantages over traditional editing methods as can be a high genome editing efficiency, a simple and easy way for its implementation and maybe the most important, the simultaneous modification at different loci of the same organism. With those advantages, the numerous modifications required for metabolic engineering approaches can be easily performed.

In the present study, a core metabolic flux model of Clostridium pasteurianum ATCC 6013 aimed to increase hydrogen production is proposed. Knockouts at the phosphoglucose isomerase; ethanol, lactate and acetate dehydronenase genes are expected in the model as potential genetic modifications in Clostridium pasteurianum. Those modifications are feasible by using the Clostridium endogenous CRISPR/Cas system as proven by Pyne et. al (2016) who worked on the same microorganism to the one considered in this study.

Via metabolic engineering and CRISPR/CAs editing, we are developing an engineered strain of Clostridium pasteurianum ATCC 6013 with a higher yield of hydrogen. This development will improve the knowledge of hydrogen-producing organisms.

Multiplexed Perturbation of Gene Expression with CRISPR Induces Epistasis and Deters Bacterial Adaptation to Antibiotics

Conference: International Conference on CRISPR Technologies

Year: 2017

Proceeding: International Conference on CRISPR Technologies

Group: Poster Session

Session: Poster Session

Time: Tuesday, December 5, 2017 - 4:05pm-6:35pm

Authors:

Otoupal, P., University of Colorado Boulder
Chatterjee, A., University of Colorado Boulder

The emergence of multidrug-resistant "superbugs" continues to pose a looming global health crisis that necessitates novel therapeutic strategies to curb their threat. The advent of CRISPR technologies has expanded our capacity to investigate such strategies. In particular, deactivated versions of the Cas9 (dCas9) enzyme enable relatively facile manipulation of gene expression in a multiplexed fashion that was once prohibitively difficult and expensive to perform.

Employing dCas9 with multiplexed sgRNAs permits manipulation of specific bacterial gene expression profiles during the critical stage of early antibiotic exposure. Previous work has consistently revealed that upon exposure to antibiotics, bacterial transcriptomes adjust to new levels that impart improved fitness phenotypes. This arises from the natural variation in gene expression across the population that exists as a bet-hedging tool to enable subsets of the population to survive during sudden environmental changes. We postulated that this heterogeneity can act as a "double-edged sword" – if a subset of random expression states provides an advantage to stress, then another subset will similarly prove disadvantageous.

Whether or not a particular expression state is advantageous or deleterious depends upon the degree of epistasis between each change. Epistasis describes the nonlinear interactions between two or more simultaneous genetic changes and is widely recognized for its role in shaping evolutionary trajectories. While research into epistasis has largely focused on simultaneous mutations, similar effects have been observed from changes in gene expression. In this study, we manipulated gene expression epistasis using multiplexed dCas9 perturbations in an attempt to curb bacterial adaption to antibiotics.

Here we present evidence that multiplexed dCas9 perturbations induce significant negative epistasis, cutting fitness of Escherichia coli up to 2-fold during antibiotic exposure. Furthermore, we demonstrate that this induced negative epistasis restricts the rate at which antibiotic minimum inhibitory concentration increases during multiple days of increasing exposure. This suggests that multiplexed dCas9 perturbations can restrict adaptation and curb the emergence of antibiotic resistance.

Prevent Vitamin D Deficiency

Conference: International Conference on CRISPR Technologies

Year: 2017

Proceeding: International Conference on CRISPR Technologies

Group: Poster Session

Session: Poster Session

Time: Tuesday, December 5, 2017 - 4:05pm-6:35pm

Authors:

Almualla, F.

Because Vitamin D deficiency is a worldwild constrain, there muat be a solution.I worked to reach the base of any problem. I got a solution which may truely help to solve the problem from its roots.

There's a known technology named as "gene therapy" this will allow us to access the gene and to reach the base.by this will can activate genes that don't form Vitamin D in places other than Kidney and Liver, as to access those it take longer time than expected.Gene therapy technology succeded previously for other disease,so why not to try it to prevent Vitamin D deficiency?

Public Understanding of CRISPR and Gene Drive Technologies and Implications for Public Acceptance

Conference: International Conference on CRISPR Technologies

Year: 2017

Proceeding: International Conference on CRISPR Technologies

Group: Poster Session

Session: Poster Session

Time: Tuesday, December 5, 2017 - 4:05pm-6:35pm

Authors:

Berube, D., NCSU PCOST GES

This paper examines the public understanding of genetic engineering, in this case CRISPR. An in-depth review of the literature and some very current findings on how the public makes sense of complicated scientific principles will be examined. The author comes from a background in the intuitive nature of public toxicological decoding when it comes to nanotechnological developments and recently extended his work in GES (genetic engineering and society). Public understanding of science and technology is affected by a broad range of heuristics and is amplified by media, esp. digital media, and the protestations from non-governmental organizations. How CRISPR is understood will impact how readily the public may be willing to accept its developments and applications. The juxtaposition of CRIPR developments and health and medicine offers a unique oppotunity to rewrite the stories on genetic engienrring. After spending two years preparing a manuscript on the ZIKV (Zika virus) and examining the options for vector control from Wolbachia approaches (MosquitoMate and Eliminate Degnue), the Oxitec engineered mosquito, and some more exotic gene drive approaches, the possiiblities of integrating these approaches into integrated pest management solution will be detailed against the body of literature on public understanding of science and technology. Lessons learned include new frames for gene drives as well as more traditional issues including receiver segmentation, message amplification, and other attenuative strategies for less than precise "false news" on gene drives. The paper concludes with a call for a marketing approach for emerging science.

Rapid Identification and Visualization of CRISPR Loci Via Automated High-Throughput Data Processing Pipeline

Conference: International Conference on CRISPR Technologies

Year: 2017

Proceeding: International Conference on CRISPR Technologies

Group: Poster Session

Session: Poster Session

Time: Tuesday, December 5, 2017 - 4:05pm-6:35pm

Authors:

Nethery, M., North Carolina State University
Barrangou, R., North Carolina State University

Clustered regularly interspaced short palindromic repeats (CRISPR) and associated sequences (cas) comprise an adaptive immune system widespread in bacteria and archaea. Located between each CRISPR repeat is a short segment of DNA collected from an invading mobile genetic element, called a spacer. Visualizing iterative spacer acquisitions representing unique evolutionary tracks has proven useful for genotyping, especially for comparative analysis of closely related organisms, and even clonal lineages. Current spacer visualization methods are tedious and typically require manual data manipulation and curation, including spacer extraction at each CRISPR locus from a genome of interest. Once spacers have been isolated, information regarding their length and content must be laboriously distilled and summarized into a format suited for comparative analysis. Here, we present a high-throughput processing pipeline and web-based visualization tool, facilitating spacer extraction, graphical comparison, alignment and clustering. The analysis pipeline automates the extraction of spacers from an unlimited number of genomes simultaneously, then feeds the resulting spacer files into a visualization engine for comparison of spacer length and content. Additional manipulation, including manual or automated multiple alignment, can be performed from the graphical user interface. This efficient high-throughput solution supports rapid analysis of large data sets and will enable and expedite large-scale genotyping efforts based on CRISPR loci.

Use of Computational Technologies to Define Potential CRISPR/Cas9 Off-Target Effects of Anti-HIV-1 Grnas in the Human Genome

Conference: International Conference on CRISPR Technologies

Year: 2017

Proceeding: International Conference on CRISPR Technologies

Group: Poster Session

Session: Poster Session

Time: Tuesday, December 5, 2017 - 4:05pm-6:35pm

Authors:

Chung, C. H., Drexel University College of Medicine
Nonnemacher, M., Drexel University College of Medicine
Wigdahl, B., Center for Molecular Virology & Translational Neuroscience
Damier, W., Drexel University College of Medicine

Despite antiretroviral therapy, human immunodeficiency virus type 1 (HIV-1) infection remains a life-long clinical problem due to reservoirs harboring proviral DNA in a latent/persistent form. Recently, gene-editing strategies utilizing the CRISPR/Cas9 system (CC9) have been developed to excise the HIV-1 genome from infected cells thereby achieving a single cell cure for HIV infection. However, due to the promiscuity of the guide RNA (gRNA) targeting, one necessary area of interest has focused on off-target activity that may cause unwanted DNA damage across the human genome. This is further complicated by the uncertainty of DNA accessibility such as the chromatin state in relevant cell populations as well as genetic variability across the human population. In order to predict the off-target effect, we have developed a screening method employing a bloom filter library constructed from a set of adapted human genomes that contains the information of all individuals within the 1000 Genomes project. In order to test the results of our method, publically available Genome-wide Unbiased Identification of Double-stranded-breaks Enabled by sequencing (GUIDE-seq) data was used from the use of HEK293T cells. Results of this analysis have shown that, in a given genomic region, the DNA accessibility implied by DNase-I hypersensitivity positively correlated to the frequency of CC9 activity. Integrating quantitative off-target data available from publications using a two-layer stacked regression model allowed better understanding of generalizable factors that contribute to off-target effect. Compiling these results will greatly increase the ability of researchers to design gRNAs that are effective with respect to targeting integrated HIV-1 proviral DNA while avoiding off-target effects.

AUTHOR INDEX